The
Universe Story

BY THE SAME AUTHORS

The Universe Is a Green Dragon
(Brian Swimme)

The Dream of the Earth
(Thomas Berry)

The
Universe Story

FROM THE PRIMORDIAL FLARING FORTH TO THE ECOZOIC ERA—A CELEBRATION OF THE UNFOLDING OF THE COSMOS

Brian Swimme and
Thomas Berry

HarperSanFrancisco
A Division of HarperCollinsPublishers

Photo credits on page ix.

THE UNIVERSE STORY: *From the Primordial Flaring Forth to the Ecozoic Era—
A Celebration of the Unfolding of the Cosmos.*

FIRST HARPERCOLLINS PAPERBACK EDITION PUBLISHED IN 1994
ISBN 0-06-250835-0 (pbk)

An Earlier Edition of This Book Was Cataloged As Follows:

Swimme, Brian.
    The universe story / Brian Swimme, Thomas Berry. — 1st ed.
        p.   cm.
    Includes bibliographical references and index.
    ISBN 0-06-250826-1 (cloth)   CIP
    1. Cosmology. 2. Civilization—History.   I. Berry, Thomas.
II. Title.
QB981.S893   1992
523.1—dc20   91-58907

06 07 08 09   RRD-H 20 19 18

To Laurance S. Rockefeller

Contents

List of Illustrations

Prologue: Botswana Storyteller, !Kung Bushman.
(*The Story*. N. R. Farbman, Life Magazine, © Time Warner, Inc.)

Chapter 1: Elementary Particle Tracks, Gluons. (Courtesy of Dr. Minh Duong-van and Stanford Linear Accelerator Center, Stanford University)

Chapter 2: Spiral Galaxy, Messier 81. (California Institute of Technology, Palomar Observatory)

Chapter 3: Crab Nebula, Messier 1. (Lick Observatory, University of California, Santa Cruz)

Chapter 4: Tropical Storm Sun, Indian Ocean. (NASA)

Chapter 5: Photosynthetic Bacteria.
(Courtesy of Dr. John F. Stolz, Duquesne University)

Chapter 6: Sperm and Egg, Sea Urchin. (Courtesy of Dr. Gerald Schatten, University of Wisconsin, Madison)

Chapter 7: Grizzly Bear, Chinook Salmon, Boreal Forest, Alaska. (Photo by Johnny Johnson)

Chapter 8: Black Bull, Lascaux Cave Paintings, Dordogne, France, c. 15,000 B.C.E. (Photo by Jean Vertut, courtesy of Yvonne Vertut)

Chapter 9: Bird Goddess, Pre-Dynastic Egypt. c. 4,000 B.C.E. (The British Museum, Department of Egyptian Antiquities, London)

Chapter 10: Guardian Kings, Longmen Caves, T'ang Dynasty, Henan Province, China, c. 675 C.E. (Asian Art Photographic Distribution, University of Michigan, Ann Arbor)

Chapter 11: Seaman's Atlas, Amsterdam, 1684. (Nederlandsche Historisch Scheepvaart Museum, Amsterdam)

Chapter 12: Albert Einstein, Pasadena, California, 1931. (The Bettman Archive)

Chapter 13: Women's resistance to deforestation, Chipko Movement, Reni, Uttar Khard Region, Himalayas, 1974. (The British Broadcasting Corporation, London)

All plates face chapter openings.

Introduction

The story of the universe has been told in many ways by the peoples of Earth, from the earliest periods of Paleolithic development and the Neolithic village communities to the classical civilizations that have emerged in the past five thousand years. In all these various circumstances the story of the universe has given meaning to life and to existence itself. The story has been celebrated in elaborate rituals. It has provided guidance and sustaining energy in shaping the course of human affairs. It has been our fundamental referent as regards modes of personal and community conduct. It has established the basis of social authority.

In the modern period, we are without a comprehensive story of the universe. The historians, even when articulating world history, deal not with the whole world but just with the human, as if the human were something separate from or an addendum to the story of the Earth and the universe. The scientists have arrived at detailed accounts of the cosmos, but have focused exclusively on the physical dimensions and have ignored the human dimension of the universe. In this context we have fractured our educational system into its scientific and its humanistic aspects, as though these were somehow independent of each other.

With all our learning and with all our scientific insight, we have not yet attained such a meaningful approach to the universe, and thus we have at the present time a distorted mode of human presence upon the Earth. We are somehow failing in the fundamental role that we should be fulfilling—the role of enabling the Earth and the universe entire to reflect on and to celebrate themselves, and the deep mysteries they bear within them, in a special mode of conscious self-awareness.

1

A new type of history is needed, as well as a new type of science. We are long past the time when history was considered to be the recorded account of the past few thousand years, and when everything prior to the Sumerian development was considered prehistory. Gone, too, is the period when the various civilizations could be explained through the sequence of their political regimes, and the listing of battles fought and treaties made. The period is also gone when we could deal with the human story apart from the life story, or the Earth story, or the universe story.

Just as surely, we are beyond the time when the scientific story of the universe could so identify the world of reality with the material and mechanistic aspects of the universe as to eliminate our capacities for that intimate communion with the natural world that has inspired the human venture over the centuries, an intimate communion that has evoked from our poets and musicians and artists and spiritual personalities all those magnificent works of celebration that we associate with the deepest modes of fulfillment of the human personality.

This new situation seems to call for a new type of narrative—one that has only recently begun to find expression. This new story has as its primary basis the account of the emergent universe such as this has been communicated to us through our observational sciences and through our capacities for reading the evidence available to us from our new instruments, with their amazing sensitivity in receiving communications from such enormous distances and through such long periods of time. We have only begun to read the immense amount of data that we now have before us. The greater problem is not in the lack of data but in our capacity to understand the significance of the data that we already possess. This data has not yet been sufficiently assimilated to bring about a new period in our comprehension of ourselves and of the universe itself.

THE MOST SIGNIFICANT CHANGE in the twentieth century, it seems, is our passage from a sense of cosmos to a sense of cosmogenesis. From the beginning of human consciousness, the ever-renewing seasonal sequence, with its death and rebirth cycles, has impinged most powerfully upon human thought. This orientation in consciousness has characterized every previous human culture up to our own. During the modern period, and

2

especially in the twentieth century, we have moved from that dominant spatial mode of consciousness, where time is experienced in ever-renewing seasonal cycles, to a dominant time-developmental mode of consciousness, where time is experienced as an evolutionary sequence of irreversible transformations.

Within this time-developmental consciousness we begin to understand the story of the universe in its comprehensive dimensions and in the full richness of its meaning. This is especially true as regards the planet Earth, a mysterious planet surely, as we observe how much more brilliant it is, when compared with the other planets of our solar system, in the diversity of its manifestations and in the complexity of its development. Earth seems to be a reality that is developing with the simple aim of celebrating the joy of existence. This can be seen in the coloration of the various plants and animals, in the circling flights of the swallows as well as the blossoming of the springtime flowers; each of these events required immense creativity over billions of years in order to come forth as Earth. Only now do we begin to understand that this story of the Earth is also the story of the human, as well as the story of every being of the Earth.

We are now experiencing that exciting moment when our new meaning, our new story is taking shape. This story is the only way of providing, in our times, what the mythic stories of the universe provided for tribal peoples and for the earlier classical civilizations in their times. The final benefit of this story might be to enable the human community to become present to the larger Earth community in a mutually enhancing manner. We can hope that it will soon be finding expression not simply in a narrative such as this but in poetry, music, and ritual throughout the entire range of modern culture, on a universal scale. Such expressions will sensitize people to the story that every river and every star and every animal is telling. The goal is not to read a book; the goal is to read the story taking place all around us.

The urgency of our time is that the story become functionally effective. The present disintegration of the life systems of the Earth is so extensive that we might very well be bringing an end to the Cenozoic period that has provided the identity for the life processes of Earth during the past sixty-seven million years. During this period life expanded with amazing fluorescence prior to the coming of the human.

But by now the human has taken over such extensive control of the life systems of the Earth that the future will be dependent on human decision to an extent never dreamed of in previous times. We are deciding what species will live or perish, we are determining the chemical structure of the soil and the air and the water, we are mapping out the areas of wilderness that will be allowed to function in their own natural modalities.

All of this is filled with risk and presumption, but if there is any way of guiding our course in such difficult decisions, it will be discovered only through an understanding of the most intimate aspects of the natural world. This is something more than our sciences are generally concerned with. A new mystique is needed, but a mystique associated with the highest level of comprehensive knowledge and critical competence.

This new period in history might be called the *Ecozoic era* to indicate the order of magnitude of the change that is taking place and of the expanded role of the human. For this new biological period to attain any degree of fulfillment will require the integral participation by all the members of the planetary community. The bioregions into which the Earth is divided in its geographical structure need to be recognized as the basis for these more integral multi-species communities. The full expression of this new orientation would bring about a movement beyond the United Nations to a United Species as the comprehensive community to which we all belong. This conception has, as its earliest manifestation, the World Charter for Nature, passed by the United Nations Assembly in 1982.

All these changes we envisage as the next phase of the larger story of the universe that is coming to expression not only in our institutions but in our language. Every change in the governing paradigm of human affairs requires an extensive change not only in our sense of reality and value but also in the language whereby we give expression to these concerns.

THIS BOOK IS written primarily for the general reader. We have kept the technical language to a minimum, and we have avoided discussing scholarly debates concerned with various episodes of the story. For instance, in our narratives of the emergence of galaxies, and of prokaryotes, and of civilizations, we employ what we regard as the most convincing hypotheses, ever aware that in the future new evidence and deeper understanding might insist that alternative hypotheses be adopted. Those readers inter-ested in more detailed discussion are referred to the annotated bibliogra-phies attached to each chapter.

This book could also provide a program for both the humanities and the science students. Indeed, this book represents a transcending of the science-humanities division in the educational process. Such a book is an urgent need for all students, since they do not currently have any adequate interpretation of the universe with which their studies are concerned. They need to give more attention to a coherent narrative of the origin and the long sequence of transformations of the universe and of the planet Earth. It is our hope that this story will provide a new unity to the educational process from its earliest beginnings through the highest level of training in the various professions. In this manner all the professions and institutions of our culture might be renewed.

The narrative of the universe, told in the sequence of its transforma-tions and in the depth of its meaning, will undoubtedly constitute the com-prehensive context of the future. Already through this story the various peoples of the Earth are identifying where they are in time and space. They are also attaining a sense of relatedness to the various living and nonliving components of the Earth community. Through this story we learn that we have a common genetic line of development. Every living being of Earth is cousin to every other living being. Even beyond the realm of the living we have a common origin in the primordial Flaring Forth of the energies from which the universe in all its aspects is derived.

"The Universe Story" refers of course to the book we have written, but only in a secondary way. The primary referent of our title is the great story taking place throughout the universe. This creative adventure is too subtle, too overwhelming, and too mysterious ever to be captured definitively. Thus we venture our telling with a certain hesitancy. Our aim is to awaken those sensitivities to the great story that enable a rich participation in the ongoing adventure. We offer this brief narrative in the hope that others will fill in what is missing, correct what is improperly presented, and deep-en our understanding of the ongoing story.

Botswana Storyteller, !Kung Bushman

Prologue:
The Story

FIFTEEN BILLION YEARS ago, in a great flash, the universe flared forth into being. In each drop of existence a primordial energy blazed with an intensity never to be equaled again. Thick with its power, the universe billowed out in every direction so that the elementary particles could stabilize, enabling the first atomic beings of hydrogen and helium to emerge. After a million turbulent years, the frenzied particles calmed themselves enough for the primeval fireball to dissolve into a great scattering, with all the atoms soaring away from each other into the dark cosmic skies opening up in the beginning time.

A billion years of uninterrupted night enabled the universe to prepare itself for its next macrocosmic transfiguration. In the depths of its silence the universe shuddered with the immense creativity necessary to fashion the galaxies—the Andromeda galaxy, the Virgo cluster of galaxies, Pegasus, Fornax, the Magellanic Clouds, M33, the Coma cluster, Sculptor, the Hercules cluster, as well as our own Milky Way galaxy—one hundred billion galaxies in all. These gigantic structures pinwheeled through the emptiness of space and swept up all the hydrogen and helium into self-organizing systems, and clusters of systems, and clusters of clusters of systems. Each galaxy presented its unique form to the universe. Each contained its own internal dynamics. Each brought forth from its own materials billions upon billions of primal stars.

The most brilliant stars rushed through their natural sequence of transformations and exploded in colossal supernovas that matched a billion stars in luminosity and spewed stellar materials throughout the galaxy. New stars formed out of the materials that had been created in the

7

billion-year processes of stellar nucleosynthesis. Second-generation stars were richer in potentiality and more complex in internal structure because the primal stars had created the elemental beings of carbon, nitrogen, oxygen, molybdenum, calcium, magnesium, and all the other hundred elements. Some five billion years after the beginning of time, the star Tiamat emerged in our spiral galaxy. Tiamat knit together wonders in its fiery belly, and then sacrificed itself, carving its body up in a supernova explosion that dispersed this new elemental power in all directions, so that the adventure might deepen.

Five billion years ago, after the universe had expanded and developed for ten billion years, our Milky Way galaxy shocked a peacefully drifting cloud of Tiamat's remnants into giving birth to ten thousand new stars. Some of these turned out to be diminutive brown dwarf stars. Others became blue supergiants that quickly flashed into the incandescence of new supernovas. Others became stable long-burning yellow stars, and still others became slumbering red stars. The universe, insisting upon diversity, also brought forth from this floating cloud of elements our own star, the Sun. Once granted existence, the Sun showed its own self-organizing abilities, blasting off nearly all the clouds of elements yet hovering about it, and spinning the rest into a multibanded disc of matter out of which arose the bonded system of Sun, Mercury, Venus, Earth, Mars, Jupiter, Saturn, Uranus, Neptune, and Pluto.

THE CHARGED EARLY planets boiled as molten and gaseous materials. On Mercury, Venus, Mars, and Pluto, chemical combinations slowly grew to become rocks and continents and planetary crust that eventually so dominated the dynamics that all significant creativity came to an end. On Jupiter, Saturn, Neptune, and Uranus chemical creativity never advanced beyond the simpler compounds, and they continued to churn primarily as gases for billions of years. On Earth, due to the balance of its own internal dynamics and its position in the structure of the solar system, matter existed as solid, liquid, and gas, and flowed from one form into another to provide an incessantly creative chemical womb from which Aries, the first living cell, arose four billion years ago.

The primal cells—the prokaryotes—had the power to organize them-selves, as did the stars and the galaxies, but they had stunning new gifts as well. The cells could remember significant information, even including the patterns necessary to knit together another living cell. Cells also possessed a new order of creativity, allowing them to fashion a chemical glove to catch the packets of energy hurled by the Sun at the speed of light, and to use these glowing quanta as food.

Spiral galaxies have the power to bring forth stars perpetually, but some of the planets that emerged from these blossomed for a time only to have their significant advances grind to a halt, as in the case of Jupiter and the other planets circling the Sun. And even in those cauldrons of creativ-ity where the universe continues to unfold for billions of years, such as the living planet Earth, all significant development can cease, as almost took place two billion years ago.

Aries and her descendents, the prokaryotes, by gathering their hydro-gen from the oceans, released oxygen into Earth's system; the oxygen slow-ly saturated the land and atmosphere and seas. By altering Earth's chemistry with this element of explosive power, the prokaryotes unknow-ingly pushed Earth's system into an extremely unstable condition, one be-yond their own capacity to endure. In time the dominant prokaryote communities perished as their interiors were set ablaze by the oxygen. But out of this crisis, threatening the very viability of the living planet, arose Vikengla, a new and radically advanced being.

Vikengla, the first eukaryotic cell, was fully capable not only of endur-ing oxygen but of shaping oxygen's dangerous energy for its own purposes, and thus seethed with creativity. The eukaryotes invented meiotic sex, and the universe's diversity expanded a hundredfold as now two genetically different beings could unite and fashion out of their genetic endowments a radically new being. Eukaryotes also invented the habit of eating living beings, and thus deepened the community of Earth not only with the inti-macy of sexual bonding but also with the intimacy associated with ecosys-temic predator-prey relationships. Finally, at the end of the period during which they were the most advanced organism in the Earth system, the eukaryotes took that daring step of submerging themselves into a larger mind as trillions of them gathered together and evoked Argos, the first multicellular animal.

Six hundred million years ago, multicellular organisms arose with a variety of qualitatively distinct body plans; they included the corals, worms, insects, clams, starfish, sponges, spiders, vertebrates, leeches, and other forms that went extinct. Life in the mesocosm had begun. Worms learned to wiggle in pursuit of soft prey, then sprouted fleshy wings to guide them through the oceans, and invented the tooth when another creature invented the shell. Ocean waves left sea plants stranded on the hot rocks; unable to crawl home they instead invented the wood cell and learned to stand up straight as lycopod trees that lived along the shores of oceans and rivers and that in turn transformed themselves into gymnosperm trees capable of covering entire continents with life. The animals followed the plants onto land, and soon the continents that had been floating lifelessly on Earth's mantle for two billion years heaved with amphibians and reptiles and insects and the great dinosaurs with gleaming eyes reaching up to the sunlit leaves of the forest canopy.

All this creativity taking place within Earth depended upon many different stabilities, including the Sun's stable burning of hydrogen, Earth's stable revolutions about the Sun, the stability of many quintillion chemical bonds throughout the Earth's system. But the galaxy is an immense home and disasters regularly visited Earth as well, most poignantly when other heavenly bodies collided with Earth and its delicate fabric of life. Sixty-seven million years ago astronomical collisions so changed Earth's atmosphere and climates that nearly all forms of animal life had to reinvent themselves or perish. Mass extinctions meant many animals followed the dinosaurs into their graves, but such destruction also opened up new possibilities, which were seized upon by the birds and the mammals, among others, who proliferated and fluoresced in the wake of the disaster.

When the mammals entered Earth's life two hundred million years ago, they developed emotional sensitivity, a new capacity within their nervous systems for feeling the universe. Throughout mammalian existence and especially during the last sixty-seven million years of the Cenozoic era, the beauty and terror of the world—the brilliance of the birds' plumage, the intoxicating display of the flowers, the lusciousness of the fruits, the frights of the forest at night, the archetypal strength of the mother-infant bond— left a deep impress on the psychic nature of all the mammals, the whales, the rodents, the sea lions, the bats, the elephants, the porcupines, the horses, the shrews, the deer, the chimpanzees, and the humans. In rare instances among the most advanced mammals, especially among the primate order, this mammalian emotional sensitivity was deepened with another neural capability, conscious self-awareness. Empowered with both, the human probed for its own distinct niche within the enveloping Earth community.

FOUR MILLION YEARS ago in Africa, humans stood up on just two limbs, and by two million years ago they began using their free hands to shape Earth's materials into tools. One and a half million years ago these restless hands were controlling fire, shaping the Sun's energy that had been stored in sticks, to advance their own projects. Beginning around thirty-five thousand years ago, as if unable to restrain any longer their astonishment at existence, humans began a new level of celebration that displayed itself in cave paintings deep within Earth, that filled the nights with festivals and music-making, that shaped ceremonials around the passing of friends and seasons, that captured in the artistic depiction of the animals some of the beauty that had seized the depths of their minds.

Twenty thousand years ago Earth, through its human element, entered conscious self-awareness of the patterns of seeds, and seasons, and the primordial rhythms of the universe. Although some of these patterns had been set into existence by Earth billions of years ago, and although the first humans had organized themselves for millions of years within these patterns, twelve thousand years ago humans began consciously shaping these patterns by domesticating plants and animals—wheat and barley and goats in the Middle East, rice and pigs in Asia, corn and beans and the alpaca in the Americas.

A secure supply of food enabled populations to surge. The first Neolithic villages to sustain human groups of more than a thousand people were Jericho, Çatal Hüyük, and Hassuna ten thousand years ago. Soon Neolithic villages arose throughout the planet as the bulk of humanity moved from its hunting and gathering mode of life into that of the settled villages, the most radical social transformation ever to occur in the human venture. In this new human context, pottery, weaving, and architecture were developed, calendars articulating the cosmic rhythms appeared, rituals and shrines to the Great Mother deity were elaborated and replaced the devotion to totemic animals of the Paleolithic. Most significantly, the great majority of the power words of the many human languages, those archetypal symbols capable of activating human genetic endowment, were established. In this period from ten thousand to five thousand years ago, the decisive developments in language, religion, cosmology, the arts, music, and dance took on their most vigorous and primordial forms, so that the urban civilizations that followed can be considered elaborations on the cultural patterns established during the Neolithic.

Five thousand years ago, the human venture mutated into a new way of life, the urban civilization. Just as the eukaryotic cells had no idea that their mutual involvements would bring forth complex animal organisms, so too those humans crowded in the Neolithic villages of Sumer had no inkling that their intensified social interactions would give rise to new power centers within the human process: Babylon, Paris, Persepolis, Banaras, Rome, Jerusalem, Constantinople, Sian, Athens, Baghdad, Tikal of the Maya, Cairo, Mecca, Delhi, Tenochtitlan of the Aztec, London, Cuzco, the Inca City of the Sun.

The invention of the bureaucratic system with its hierarchical authority relations and its emphasis on specialization made possible vast transformations of the human and the natural processes. Rivers became irrigation systems for plowed fields. Commercial transactions engaged the energies of entire nations as caravans crisscrossed the world and forests were changed into shipping enterprises. Population and wealth soared, pyramids rose up along with elaborate temples, ornate shrines, lavish palaces, and cathedrals. To protect these urban centers with their concentrations of wealth and power and to maintain great territories of the planet under the rule of a state-promulgated legal enactment, military establishments came forth, with their weaponry and fortifications and chronic warring processes crossing entire continents, supported by a cast of warring deities that had replaced the image of the Great Mother as the principal symbols for the human enterprise.

In the middle of this turbulence, the pathos of the human condition and the promise of a transcendent realm beyond the pathos—the Tao,

Brahman-Atman, Heaven, Nirvana—impressed themselves upon the human mind. There arose the universalist beliefs of Buddhism, Christianity, and Islam, which came to pervade the planet's centers of civilization from Europe across North Africa and India and throughout the Eurasian continent to China and Southeast Asia. Only sub-Saharan Africa, the Americas, Australia, and pockets of indigenous peoples entirely escaped the control and influence of these four civilizational complexes, the Middle East, Europe, India, and China.

Five hundred years ago, Europeans initiated the third of humanity's great wanderings. The first had brought Homo erectus north out of Africa to spread throughout Eurasia. On the second, Homo sapiens wandered until they reached the Americas and Australia. The principal difference in the modern break-out of the sixteenth and seventeenth centuries was that now Europeans encountered humans wherever they went; equipped with superior technologies and bureaucratic social systems, they colonized peoples all around the planet, especially in the Americas and Australia. In the nineteenth century, India was added as a colony, and Japan and China were forced into trading patterns with the European enterprises. The political-cultural shape of humanity was thus altered in a radical manner as the various human communities were in contact with each other and turned toward a common destiny in a way never previously existing.

While these global political connections were taking shape, Europe's own internal articulation came in the form of the nation-state with its self-government. This liberal democratic movement, which would spread throughout the planet, had its violent beginnings in the American Revolution of 1776 and the French Revolution of 1789. Throughout the nineteenth and twentieth centuries, the nation-state provided the integrating community, replacing the former contexts of the band, or the village, or the capital city with its surrounding territory. The sacred mystique of the nation-state could be found in the ideals of nationalism, progress, democratic freedoms, and individual rights to private property and economic gain. Thus conflicts between nation-states took on the character of holy wars over these sacred ideals, culminating in the intra-European tensions that soon engulfed the whole world of humanity during the twentieth century.

The dominant entity that emerged was not any particular nation-state but the multinational corporation. These new institutions directed vast scientific, technological, financial, and bureaucratic powers toward controlling Earth processes for the benefit of the human economy. By the end of the twentieth century, the destruction left by the wars between nations was dwarfed in significance by the destruction of the natural systems by industrial plunder. In geological terms, human activities in the twentieth century ended the sixty-seven-million-year venture called the Cenozoic era.

As INDUSTRIAL HUMANS multiplied into the billions to become the most numerous of all of Earth's complex organisms, as they decisively inserted themselves into the ecosystemic communities throughout the planet, drastically reducing Earth's diversity and channeling the majority of the Gross Earth Product into human social systems, a momentous change in human consciousness was in process. Humans discovered that the universe as a whole is not simply a background, not simply an existing place; the universe itself is a developing community of beings. Humans discovered by empirical investigation that they were participants in this fifteen-billion-year sequence of transformations that had eventuated into the complex functioning Earth. A sustained and even violent assault by western intelligence upon the universe, through the work of Copernicus, Kepler, Galileo, Newton, Buffon, Lamarck, Hutton, Lyell, Darwin, Spencer, Herschel, Curie, Hubble, Planck, Einstein, and the entire modern scientific enterprise, had brought forth a radically new understanding of the universe, not simply as a cosmos, but as a cosmogenesis, a developing community, one with an important role for the human in the midst of the process.

Over fifteen billion years the universe brought forth stars, galaxies, supernovas, the first cells, the advanced eukaryotes, the proliferation of the animals and plants, and the conscious self-awareness that has come to permeate so thickly the many components of the Earth community. The future of Earth's community rests in significant ways upon the decisions to be made by the humans who have inserted themselves so deeply into even the genetic codes of Earth's process. This future will be worked out in the tensions between those committed to the Technozoic, a future of increased exploitation of Earth as resource, all for the benefit of humans, and those committed to the Ecozoic, a new mode of human-Earth relations, one where the well-being of the entire Earth community is the primary concern.

Elementary Particle Tracks, Gluons

1.
Primordial Flaring Forth

ORIGINATING POWER BROUGHT forth a universe. All the energy that would ever exist in the entire course of time erupted as a single quantum—a singular gift—existence. If in the future, stars would blaze and lizards would blink in their light, these actions would be powered by the same numinous energy that flared forth at the dawn of time.

There was no place in the universe that was separate from the originating power of the universe. Each thing of the universe had its very roots in this realm. Even space-time itself was a tossing, churning, foaming out of the originating reality, instant by instant. Each of the sextillion particles that foamed into existence had its root in this quantum vacuum, this originating reality.

The birth of the universe was not an event in time. Time begins simultaneously with the birth of existence. The realm or power that brings forth the universe is not itself an event in time, nor a position in space, but is rather the very matrix out of which the conditions arise that enable temporal events to occur in space. Though the originating power gave birth to the universe fifteen billion years ago, this realm of power is not simply located there at that point of time, but is rather a condition of every moment of the universe, past, present, and to come.

PARTICLES, LIGHT, AND time emerged in the beginning. Space, too, unfurled out of potentiality and has continued to unfurl each instant of cosmic existence. In the beginning space foamed forth to create the vast billowing event of the expanding universe. The universe venture was under way. Had the originating powers not gushed forth a world-creating space and time, our cosmos would have existed a quintillionth of a second, just a pinprick event that would have instantly snuffed itself out. The cosmic adventure of fifteen billion years has depended upon the ever-fresh unfurling of nascent space.

The rate of the spatial emergence reveals a primordial elegance. Had space unfurled in a more retarded fashion, the expanding universe would have collapsed back into quantum foam billions of years ago. Such a collapse would have taken place even if space had unfurled one trillionth of a percent more slowly. If space had emerged more rapidly, equally disastrous results would have followed. The constituents of the universe would have been too widely separated for anything truly interesting to happen.

The original body of the universe maintained itself in a delicate balance. If either the rate of spatiation or the power of gravitation had wavered too far one way or the other, the adventure of the universe would have ceased. For instance, the universe would never have reached the moment when living cells sprouted forth. The vitality of a dolphin as it squiggles high in the summer sun, then, is directly dependent upon the elegance of the dynamics at the beginning of time. We cannot regard the dolphin and the first Flaring Forth as entirely separate events. The universe is a coherent whole, a seamless multileveled creative event. The graceful expansion of the original body is the life blood of all future bodies in the universe.

Though this law of expansion was fixed from the earliest instant of existence, other laws were not yet formed at that time. The first particle interactions were not fixed and determined in the way they are today. There was an element of freedom, of randomness, associated with these interactions. The electrons, positrons, the quarks, the neutrinos had not yet achieved their identity. They enjoyed a chaotic freedom of possibilities they would later be denied. Most likely there were even from the beginning innate biases for certain kinds of intensities to these interactions, but there were also in every interaction degrees of freedom that would disappear in later eras.

The first epoch of the Flaring Forth reached its end when the freely symmetric interactions hardened into a structure. Suddenly the universe as a whole changed phases. What had been symmetric and free was now fixed into particular interactions with determined intensities—the gravitational, the electromagnetic, and the two nuclear interactions.

These four laws theoretically could have been very different: different in number, intensity, character. Why did these particular four emerge? Perhaps their final form even depended to some extent on the experimentation and exploration of the former, freer era. Perhaps their structure was determined to some degree by what had preceded the moment of symmetry, when a pure or at least original activity had settled into a particular fixed form. If so, these four interactions can be regarded as analogous to habits that the universe adopted for its primary actions. Thus the one primordial act of the universe now appeared as four different activities.

In this phase transition the fundamental architecture of the universe's interactions was set for all time. It was not yet certain where the largest stars would appear, but the upper limits to their sizes and intensities were already fixed. It was not yet certain how many planets would come into existence, but an invisible ceiling for their highest mountains was already in place because the strengths of the interactions of the mountains' constituents were now in place. It was not at all certain if bivalve mollusks would ever exist, but the possibilities for shell sizes were now determined. It was certainly far from obvious whether or not there would ever exist anything like a mammal, but the fundamental range for how high they could leap or how powerfully they could clamp their jaws was now set into the sinews of the universe.

The universe established its fundamental physical interactions in a manner similar to the way it unfurled its space—with stunning elegance. Had it settled on a slightly different strong interaction, all the future stars would have exploded in a brief time, making an unfurling of life impossible. Had the universe established a slightly different gravitational interaction, none of the future galaxies would have taken shape. The integral nature of the universe is revealed in its actions. The universe as it expands itself and establishes its basic coherence reveals the elegance of activity necessary to hold open all the immensely complex possibilities of its future blossoming.

I N THE PRIMORDIAL fireball that followed the creation of the laws, nothing was sufficient unto itself. Just as the interactions were established, the latent heat released by the phase transition generated a vast storm of particles, which glimmered briefly, then dissipated, to be replaced by a new world of particles. Each thing was an emergent thing, a concentration of an energy that had been given, a contingent evanescent being that had only recently appeared in the world of existence, freshly, for the first time. Each thing in the great Flaring Forth existed for only the briefest of instants. For in the beginning, both before and after the fundamental laws were established, nothing was permanent. All the particles—the quarks, the electrons, the protons, the muons, the photons, the neutrons, as well as their antiparticles—would cascade into existence, interact with other particles, then disappear. Where did the particles go? Into the same night that had given them forth, into nonexistence, absorbed back into that abyss, that originating and annihilating power that is the marrow of the universe. If some proton narrowly missed colliding with a muon, thus clinging mysteriously to existence another instant, if it miraculously missed another ten million positrons, neutrons, and anti-protons, and thus continued soaring in its wild zigzag dance at the beginning of time, in another instant it would inevitably encounter some particle and both would disappear from existence, their brief bright sojourn dissolving forever into that great power that had accompanied them each instant of the journey.

In the beginning was a flashing forth of evanescent beings. In every instant the universe was fresh, just as a flickering flame's shape is fresh, newly created. In the beginning the universe was a sparkling. Nothing endured the beginning except the flickering creativity bringing forth each new billowing. The intensity, the concentration, the shimmering of the beginning was so extreme that no single being in the entire universe endured it, but each thing disappeared almost as suddenly as it entered existence.

The image of a rapidly expanding ball of red fire fails to capture an extremely important fact of the beginning: there was no outside. One cannot imagine the Flaring Forth as taking place off in some distance that we can then view. Each point of the universe was in the fire, was immersed in the billowing forth of the universe. To ask, "If we were in the fireball, what would we see?" fails to appreciate the extreme nature of the event. Any carbon-based life would instantly volatilize if placed in the beginning.

We live in a world of green maple leaves, of cirrus clouds brushed in dry strokes on a darkening blue sky, a world where sea gulls shriek over the entrails thrown from the fisherman's home-bound trawler rocking and yawing in the great ocean drifts, the half-moon lifting above the horizon. Our senses and our imagination have been fashioned here. Our bodies and their sensitivities are home here, and yet all this world has for its direct ancestry an event whose dimensions break outside all the experiences humans have had in two million years of existence. In that primordial reality the greatest of the Himalayan mountains would dissolve more suddenly than would a child's sand castle hit by a tsunami wave. The Earth's solidity becomes smoke in the beginning. In that beginning time, the briefest human reverie, an unnoticed flicker of a mind on a summer's day, would be an interval of time in which the primeval fireball thundered through a thousand universe annihilations and as many universe rebirths.

At the base of the serene tropical rainforest sits this cosmic hurricane. At the base of the seaweed's column of time is the trillion-degree blast that begins everything. All that exists in the universe traces back to this exotic, ungraspable seed event, a microcosmic grain, a reality layered with the power to fling a hundred billion galaxies through vast chasms in a flight that has lasted fifteen billion years. The nature of the universe today and of every being in existence is integrally related to the nature of this primordial Flaring Forth. The universe is a single multiform development in which each event is woven together with all others in the fabric of the space-time continuum.

After less than a single second—but already an interval of time in which the universe had transformed itself many millions of times—the second great macrotransition of the cosmos began. The universe had expanded to the point where the energy of the photons was no longer capable of evoking new particles from the quantum vacuum, that realm of cosmic fecundity. The kaleidoscopic sparkling between existence and nonexistence was coming to an end. Now when particles encountered their antiparticles and disappeared into light, the photons brought forth no new set of particles. The total number of particles was dropping steadily.

When this mass annihilation ended, only one billionth of the original matter in the universe remained. This tiny sliver of the primordial universe managed to slide through this eye of the needle near the beginning of time and thus entered a new state of being. The particles that survived the great annihilation could now endure.

Henceforth, instead of having its existence vanish with any interaction, a particle could persist and could even enter enduring relationships. A proton and a neutron joined together. Two protons and two neutrons, two protons and one neutron. Such primal enduring partnerships entered existence for the first time. The first stable ground of the universe made its appearance. All future ground—whether that of stars or of planets or of continents—would find its strength from this, the world's first foundation.

In this macrotransition, too, the elegance of the universe's power to unfurl space shows itself. If the unfurling had been somewhat slower, so that the temperature of the universe dropped more slowly, the "window" enabling nuclear particles to enter bound relationships would have remained open longer. The protons and neutrons would not have stopped at helium or lithium, but would have continued gathering together until they formed iron nuclei. If such had happened the adventure of the universe would have been reduced to the ever wider dispersion of inanimate iron atoms. But instead, the universe maintained itself on the edge of a knife—expanding in its delicate fashion so that in the beginning the lightest nuclei could stabilize, nuclei whose powers were essential for the emergence of the first living cells.

WHAT CAN WE mean when we say, "The universe maintained itself?" Is "universe" meant as an agent of activity? Does "maintain" indicate foresight on the part of such an agent? And by "knife edge" do we mean to imply that things in the universe could have been otherwise? Or are things determined to be as they are, and is the notion of "knife edge" really an illusion?

Traditionally, the cosmological enterprise aimed at an understanding of the universe and the role of the human in the universe. But over the last three centuries cosmology has come to mean "mathematical cosmology," the search for empirically based answers to a core set of questions such as How big is the universe? How old is the universe? What is the universe made of? How did its structures evolve? How long will these structures last? Each of these questions focuses on a different aspect of the physical universe and has, at least theoretically, a mathematical answer or set of answers that satisfies the query. Traditional questions concerning the role and meaning of human beings were thus relegated to others so that the scientific enterprise could concentrate without distraction on the physical facts of the matter.

So long as the vast universe could be considered "out there," this separation of investigations was a reasonable procedure, for the concerns and feelings of tiny humans on an obscure planet seemed to have only negligible connections with the great immensity of the physical universe. But this very concentrated study of matter concluded that the universe is not just a vast "out there," but is rather an "in here." Numinous fire became, over fifteen billion years of creativity, the here and now—a moving endpoint of development, one that happens to include communities of living beings. It was this very scientific enterprise that articulated the connections between the existence of life forms seeking a way to live a worthwhile life, and the dynamics at the beginning of time.

Suddenly it has become clear that at least the scope of the questions common in traditional cosmology must be synthesized with the factual investigations of scientific cosmology.

Mathematical cosmologists look about themselves and see the stars and galaxies and ask, "What is the nature of the fireball that could enable the development of these structures?" Out of the very discoveries they have made, we are now emboldened to extend such questioning: "Given the existence of mountain wildflowers, what is the nature of the Flaring Forth at the beginning of time? Given Mozart's symphonies, what is the nature of the dynamics of the universe that could have led to such structure? Given the care with which a mother lark will nurture and protect her young, what is the universe made of? Given the direct influence humans have on the functioning of the planet, what are the long-range consequences human activity will have on cosmic evolution?"

Cosmology aims at articulating the story of the universe so that humans can enter fruitfully into the web of relationships within the universe. Seen within the entire history of the universe, this enterprise could hardly be more traditional, for relationships are regularly created, explored, developed, ended, and reinvented at every level of being. The establishment of the four fundamental interactions illustrates this activity at the particle level, and later we will see it for other episodes of the universe story. Indeed, the cosmological enterprise is at a pitch of intense creativity in our time precisely because the role of the human in the web of relationships is changing so radically.

To articulate anew our orientation in the universe requires the use of a language that does not yet exist, for each extant language harbors its own attitudes, its own assumptions, its own cosmology. Thus to articulate anew the story of our relationships in the world means to use the words of one of the modern languages that implicitly, and to varying degrees, obscures or even denies the reality of these emerging relationships. Any cosmology whose language can be completely understood by using one of the standard dictionaries belongs to a former era.

The cosmologies adequate for our present challenge of reinventing the human will necessarily need to reinvent language to some degree. The definitions in the dictionaries of the future will in essence refer back to those cosmological ventures of our time that humans come to embrace as adequate articulations of their roles and their relationships in the surrounding story. For this reason alone, an encounter with the new cosmology is a demanding task, requiring a creative response over a significant period of time. Human language and ultimately human consciousness need to be transformed to understand in any significant way what is intended.

What then can we mean when we say, "The universe maintained itself on a knife edge?" To begin an explanation we need to probe the nature of gravity. What is gravitation? For the classical mechanistic understanding, the word *gravitation* simply indicates a particular attraction things have for each other. In Newton's theory this is called *force,* in Einstein's theory it is called the *curvature of the space-time manifold.* Basically, humans noted the strong bond the Earth had for all its materials. The Sun and the Moon and the stars seemed free of this bond until Newton convinced the intellectual community that Earth and Sun and all the stars were as bonded as Earth and the falling apple. As our powers of observation increased we came to realize in our own century that there were billions of galaxies entirely outside our Milky Way galaxy and that all of these too were bonded together by a primordial attracting power permeating the universe. But even now, after Newton and Einstein and others, we have not moved closer to saying what gravitation is in itself.

The gravitational bond holding each thing in the universe to everything else is simply the universe acting. We call such a bond primordial or fun-

24

damental because there is nothing else in the universe that is beneath this activity, nor behind it, nor above it, nor inside it. The gravitational bond is called primordial because whatever it is, it is an illustration of pure activity—simple, original, foundational.

To say the stone falls to the Earth misses the active nature of the event. To say gravity pulls the stone to the Earth suggests an underlying mechanism that has no basis in reality. To say the Earth pulls the rock to itself fails to capture the mutual presence of the universe to each of its parts. It is more helpful to say that the planet Earth and the rock are drawn by the universe into bonded relationship. The bonding simply happens; it simply is. The bonding is the perdurable fact of the universe and happens primevally in each fresh instant, a welling up of an inescapable togetherness of things.

I T IS DIFFICULT to entertain the phrase "the universe acting" within modern consciousness. Our inability reflects the "pluriverse" thinking of the modern period. By pluriverse we mean the tradition of dividing activity into fundamentally distinct subactivities. Gravitation was considered separate from electromagnetism and both separate from the second law of thermodynamics. The tack of scientific exploration was to divide, specialize, and abstract. No one had any hope for deeper understanding unless attention were restricted in this way to a particular segment of universe activity. If a group of brilliant humans focused their knowledge and imagination upon gravitational attraction, positive advance in understanding could be expected. And indeed that is what happened. Thus it came to be that all knowledge was divided into that pertaining to gravitation, or that pertaining to the strong nuclear interaction, and so forth.

Surprises were in store for this enterprise. The work of James Clerk Maxwell (1831–1879) announced a change of tide. Maxwell realized that the two interactions called electricity and magnetism were in fact not two. He showed with his mathematics that magnetism and electricity were both abstractions of a more fundamental reality, electromagnetism. Under particular circumstances electromagnetism appeared to be just electricity, under others just magnetism, but in fact electricity considered as separate from the electromagnetic interaction was an intellectual abstraction without physical reality.

Albert Einstein (1879–1955) was so deeply convinced that Maxwell's move had significant implications for our understanding of the universe that he devoted the greatest bulk of his creative scientific career to pushing this line of investigation. And although his achievements in this line of investigation were ambiguous, his intuition was deeply affirmed when physicists demonstrated that electromagnetism and the dynamic involved with radioactivity, the weak nuclear interaction, were in fact not two. They were instead different manifestations of a more fundamental ordering now called electroweak.

Some of the most exciting and absorbing work in theoretical physics today involves the investigation into the unified nature of all the interactions. These "theories of everything" express the scientific insight that reality is a universe and not a pluriverse. We used this line of thinking in narrating the first era of the Flaring Forth when a unified and free interaction constellated into the fixed laws of the physical universe.

Although we today are just beginning to realize what it means to say that the universe acts in an integral manner, future humans will take this for granted. They will begin with an understanding that there are not four fundamental interactions, nor five, nor six, nor any higher number, but that all interactions are different manifestations of primordial universe activity. The strong interaction holding the nuclei together is distinct from electromagnetism and will remain so; but the two are not finally separate. Each is a particular form of something more basic. In this cosmology we say simply that the strong nuclear interaction is the universe acting; the gravitational interaction is the universe acting; the thermodynamic dynamism toward entropy is the universe acting.

More important yet and more difficult to grasp, the gravitational interaction, even after it has split from the others at the symmetry breaking, never happens as a separate activity. "Gravity" is an intellectual abstraction of tremendous significance and power; but the universe acts as a whole in each case, and never as gravity by itself. When we consider a rock falling to the earth and calculate the gravitational geodesic, we do so by ignoring the electromagnetic interaction. We are justified in doing so because of the negligible effects electromagnetism has on the falling rock. But the practical considerations of calculating the rock's trajectory do not alter the actual fact of what takes place. As any rock falls, the electromagnetic interaction and the weak nuclear interaction are all in act, are affecting things. It is never the case that gravity works alone while the other interactions slumber. Gravity is not a being that acts; it is not an independent power that acts. Nor are the other interactions beings who act. Nor is electricity an independent power. Always and everywhere, it is the universe that holds all things together and is the primary activating power in every activity.

THESE LAST PARAGRAPHS were devoted to explaining the sense in which "universe" is used as the subject of a sentence. We can now develop this way of thinking with the assertion that the universe is not a thing, but a mode of being of everything. Every being has its particular mode of being and its universe mode of being, its "microphase" mode and its "macrophase" mode.

To illustrate the meaning in these assertions, consider a proton in the great cascade of happenings during the primeval fireball a hundred thousand years after the universe's birth, when the cosmos had expanded to a million light years in diameter. What do we need to take into account to tell the story of the proton? During all three centuries of the modern scientific enterprise the answer was clear. We could ignore everything but a small region surrounding the proton. A little circle drawn about its place in space-time was enough to understand it. The entire past, present, and future could be put to one side. The **nature** of the proton—by which we mean its physical parameters—could be described by concentrating on that single tiny bubble.

The assumption that a thing's nature is determined by its immediate surroundings is equivalent to the assumption that a thing's full reality is contained in its immediate vicinity. Such an assumption has proven of great value in the history of science. But from the beginning, questions concerning the validity of this assumption have persisted. Even Isaac Newton's theory had built into it features suggesting that the presence of a particle was co-extensive with the universe. Such an implication went beyond what either Newton or his contemporaries were ready to accept. In the end Newton threw up his hands, so to speak, and said he wouldn't interpret the meaning of his equations. He only claimed that they gave accurate results.

Some of history's most spectacular minds wrestled with this question of a thing's location or presence, including Newton and Leibniz in the seventeenth century, and, in the quantum context of the twentieth century, Albert Einstein and Niels Bohr. One great difference separated the early debates from those of our century. Three centuries of scientific discoveries had produced instrumentation of extremely subtle sensitivities. In the 1970s technological capabilities were in place to allow Alain Aspect in France and John Clauser in the United States to put the debate to the empirical test. In these and other repeatable experiments scientists established that it is not viable to think of a particle or an event as being completely determined by its particular locale. Events taking place elsewhere in the universe are directly and instantaneously related to the physical parameters of the situation.

Our tentative interpretation is that if one seeks to understand the nature of a proton in a particular region of the primeval fireball, it is not enough to take into account only the immediate region of space-time. Other events are involved. Conversely, to understand the full presence of a proton it is not enough to describe its effect in the local region, for distant regions of the fireball are directly correlated to what takes place with the proton. This connection is instantaneous; it is not mediated through space and time. For this connection to be immediate, there must be a way in which the events of the space-time manifold are connected beyond space and time.

We are only at the beginning of this exploration in science. Conclusions and interpretations must be made tentatively. Certainly common sense advises us to consider the influence of distant events as being minimal, even negligible, in all but a few rare situations. But practical considerations are beside the point here. At issue is not the physical magnitude of such effects; at issue is the correct understanding of a proton's nature.

The counterintuitive conclusion we draw from this line of investigation is that to speak of a proton as a separate particle restricted to a certain patch of space-time is to speak of its microphase mode of being, a valid though limited understanding. The macrophase mode or presence of the proton includes all particles with which it is correlated, which includes all those particles it has interacted with at any time in the past. Since the universe bloomed from a seed point, this means that a full understanding of a proton requires a full understanding of the universe. The fireball manifests itself as a quintillion separate particles and their interactions, but the nature of each of these particles speaks of the universe as indivisible whole. No part of the present can be isolated from any other part of the present or the past or the future.

The story then of a single proton is, in the sense indicated, integral with the story of every other particle in the primeval fireball. To tell the full story of a single particle we must tell the story of the universe, for each particle is in some way intimately present to every other particle in the universe.

THE UNIVERSE BLOOMED into existence, settled on its fundamental laws, and stabilized itself as baryons and simple nuclei. For several hundred thousand years it expanded and cooled and then, in an instant, at the very end of the fireball, the universe transformed itself into the primordial atoms of hydrogen and helium. Our wandering proton snapped into a new relationship with one of the erstwhile freely interacting electrons. These bonded relationships were impossible during the violent former eras, but now they became the predominant mode of reality.

The creation of the atoms is as stunning as the creation of the universe. Nothing in the previous several hundred thousand years presaged their emergence. These dynamic twists of being leapt out of the originating mystery and immediately organized the universe in a fresh way. Is it the electron trapping a proton? Or vice versa? Or is it the "electromagnetic interaction" trapping electron and proton?

It is rather an event initiated by the universe, and completed by the mysterious emergent being we call hydrogen, a new identity that has the power to seal a proton and an electron into a seamless community.

Atoms of hydrogen and helium formed—seemingly such microcosmic events—and yet the fundamental qualities of the fireball were changed forever. Hydrogen and helium allowed light to shoot through them. The universe broke itself apart to begin an entirely new era of the macrocosmic adventure, an adventure opened up by the creativity of these first hydrogen and helium beings.

29

Spiral Galaxy, Messier 81

2.
Galaxies

O UT OF QUANTUM chaos, the great power of the Flaring Forth established its fundamental laws and its first stable foundations. As space unfurled within the fireball, a subtle and seemingly insignificant perturbation appeared in what would otherwise have been a perfectly smooth and uniform event. The density of the matter and energy in the fireball fluctuated slightly. That is, slightly greater concentrations of the emergent energy began to form. Had sentient beings been there they would not have even noticed this fluctuation. Even so, in the journey of this universe, events that begin in innocuous ways often grow into stupendous power. Only because of these quantitatively minor waves would galaxies or stars or planets or life ever emerge in future time.

How would extremely slight fluctuations become determining realities in our universe's development? Originating power that evoked the cosmic seed from the generative potentiality at the beginning of things was rapidly infusing the universe with space, thereby transforming microcosmic events into macrocosmic realities. Quantum fluctuations such as those taking place incessantly throughout the foam of space-time were not allowed to dissipate back into nonexistence, as is usual in our time, but were suddenly made to serve as the cosmic patterns for all future existence. In a flash the specific, accidental, and random shapes of these particular quantum bursts zoomed to giant size. The most insignificant ripples were suddenly vast realities that set contours into the cosmic adventure, because these ripples gathered matter and energy to themselves slightly faster than the universe expanded.

Rather than uniform matter and energy spread throughout space-time, the universe presented itself with fluctuations in its fundamental constituents.

31

The action is similar to the way the waves in the ocean will trace a signature of ripples in the beaches of every continent. From far above, the beaches appear perfectly flat, leveled each second by the waves. Only close inspection by bare feet can reveal the delicate tracings of rhythms whose source is in the complex wave action along the beach. This wave action makes present the gravitational rhythms of Earth and Sun and Moon with their connections to all the trillions of stars. So this analogy between two vast and complex systems etching minute ripples on an underlying fabric is clear enough to indicate how the ripples form. But the analogy ultimately fails because these ripples on the beach do not become determining realities for the future evolution of the ocean. Such would be needed to capture the marvel of the situation where slight ripples gave birth to the galaxies.

In order for the ripples of the primordial Flaring Forth to shape the universe in its cosmic form, the primordial atoms were required. As the fireball cooled, in a single instant, throughout all the trillions of cubic miles of the extant universe, hydrogen and helium emerged. Hydrogen and helium are dynamic centers of action. If we are to understand the manner in which the universe fabricated itself into galaxies, we need to probe the way hydrogen and helium transformed the nature of the large-scale universe.

THE POWER AT THE beginning of time drew forth a universe, unfurling vast reaches of space within itself. This power then articulated itself in a diversity of power centers at the level of atomic constellations.

Hydrogen is not inert, dead, or passive matter. Rather, each hydrogen atom seethes with its own particular energies, instant by instant. The hydrogen atom is an accomplishment requiring constant communicative action among its constituent parts—the proton, the electron, and the photons. That power, that dynamic center of activity, that identity that seals electrons, protons, and photons into a coherent whole, is hydrogen. Hydrogen is this new power, this new center of activity, this new presence in the universe. It is a new actuality, a thing that acts in a new way.

Hydrogen emerges out of primordial universe activity, but it has its own distinctive modalities. An electron when alone will interact with a particular kind of photon, but this same electron in a hydrogen atom will not interact with that photon. An electron's interaction with the wide universe differs, depending on whether or not it is within a hydrogen atom. Within the hydrogen atom a new mode of presence is established.

The interactions between photon and hydrogen, between hydrogen and hydrogen, or between hydrogen and helium are all new. The laws constituted among the elementary particles continue, but a new weave of order emerges with the creation of these primordial atoms. Hydrogen does not enter the universe as a new thing in an established and fixed world; rather, hydrogen enters and weaves a new universe. The emergence of the primordial atoms constitutes a recoding of the universe, a recoding that includes the order established by the elementary particles, but now in a cosmos whose large-scale structure is newly created.

For most humans, sensory contact with helium is limited to those bright balloons in the shape of Mickey Mouse that festoon our carnivals, sometimes floating away from a child's hand and bringing tears and grief as they lift into the blue sky. The contact our eyes establish enables us to assume easily that helium is an invisible, inert gas—just this stuff, passive, just there. In actuality each helium atom roars with activity. In the time it takes a human to sneeze, a single helium atom has had to organize a billion different evanescent events to establish its helium presence in the world. Just one of its accomplishments is to keep its electrons free from interacting with most of the photons rushing at it. To exist as an invisible gas is a major achievement, one requiring instant-by-instant action, an accomplishment that transformed the universe.

In the pre-helium and pre-hydrogen world, the universe's storm of energetic photons would travel at light speed only an instant of time before ramming into, scattering, transforming, or demolishing another particle, then to repeat a similar act in the next instant. But with the birth of the primordial atoms, all the photons shot off forever in whatever direction they happened to be traveling the moment before, flying quickly and silently through these new clouds of beings, traveling right through them. Hydrogen is invisible because hydrogen assembles itself in a way that allows photons to rush right through it without ever meeting an obstacle.

The universe became transparent. Suddenly the pressure that had been exerted by the photons for the last million years vanished. Suddenly the formerly insignificant ripples in the fireball could flex their muscles and grab matter to themselves. Suddenly the universe constellated into a trillion separate clouds of hydrogen and helium. New presence emerged; powers of self-determination erupted within each of these clouds. The galaxies were born.

The universe, by expanding itself and by transforming micro-ripples of space-time into macro fluctuations that fractured the universe into a trillion parts, initiated the birth of galactic clouds. The universe in this sense was a power that brought into existence clouds of hydrogen and helium, but once these clouds were established, their own self-organizing dynamics came to dominate the evolutionary adventure.

The power that evoked the universe evoked a new power in the form of the galactic cloud. This dynamic of a power evoking beings with new modes of power happened both in the birth of the primordial atoms and in the birth of the galaxies and is a fundamental theme throughout fifteen billion years of cosmic development.

The galaxy set to work immediately, sweeping vast reaches of itself into a central concentration of matter and energy so extreme it punctured the very fabric of space and time. These "black holes" at the galactic centers spun their immense masses at revolutions faster than a thousand times each second. Density waves, initiated by the galaxy as a whole, swept through the galaxy's clouds. These new ripples amplified, damped out, and sometimes superseded the fluctuations that had been generated by the universe as a whole. These new density waves shocked clouds of hydrogen and helium to condense rapidly into thousands of stars at a time.

After a billion years, the universe broke itself into a trillion clouds, each with its own dynamics, enabling each cloud to escape from the universe's expansion in that the diameter of each cloud remained the same while the space between the clouds was progressively extended. The galaxies collapsed hydrogen clouds into stars and the universe once more burst into radiance. Galactic formation was most intense on the surface of great cosmic bubbles housing cold dark matter within. After the fireball ended, the universe's primordial blaze extinguished itself, only to burn once more in the form of millions of lacy veils of galaxies filling all one billion light years of space-time existence.

On one of these gossamer filaments hung the Virgo cluster, a self-contained gathering of a thousand separate galaxies. Bonded to this cluster were a great many clusters, including one pinwheel formation of two dozen galaxies having two poles: the Andromeda galaxy on one side and our own home, the Milky Way galaxy, on the other. Within the Milky Way, we live in a stellar system twenty-eight thousand light years from the galactic center, two-thirds the distance to the edge. Our Sun is one of the hundred billion stars that the Milky Way swirls around itself, each of them spinning about in a bonded relationship with every other one. And the Milky Way remains bonded to all one hundred billion galaxies of the cosmos, for instant by instant the universe creates itself as a bonded community.

Perhaps enough has been said to indicate our meaning in using *universe* as the subject of a sentence. But we need now reflect on the use of the word *bonded,* with its connotations of emotional intimacy and animal feeling. Is there any sense or justification for using such emotionally charged language when speaking of an event that took place billions of years before the existence of animal emotions?

From the perspective of the worldview that dominated before the discovery of the universe as story, all such "human" language, when used to describe the universe, was regarded as anthropomorphism, the projecting of inner human feelings onto what was understood to be an inert, unfeeling universe. Every effort was made to scour language free of such anthropomorphic stains so that the words used would be neutral. Best of all, of course, was mathematics. The equations of mathematical physics became an ideal on which all human language about the universe should model itself.

From such a viewpoint, any use of a word like *bonded* to describe what took place in the galaxies would be considered unjustified, misleading in its connotation of human bonding, something to be avoided at all costs. In more specific terms, the gravitational interaction between a proton and a neutron or the electromagnetic interaction between an electron and a proton were entirely distinct from what took place between a female goose and her bonded mate. To use language appropriate for the geese when describing protons would be to distract from that protonic reality with language that is unrelated, distant, irrelevant.

This program to cleanse language of its specifically human content was born out of a desire to establish contact with the things themselves, with their nature, with the very essence of what was out there confronting us. The western mind had become completely fascinated with the physical dimensions of the universe. It would establish direct contact no matter what the cost. A language stripped of all anthropomorphism was sought. A language free of all human wishful thinking and all human illusion was the ideal. A univocal language was needed, one whose words were in direct, one-to-one relationship with the particular physical aspects under consideration. In this way, anthropomorphic language was abandoned in favor of mechanomorphic language.

The exemplar in this regard is Isaac Newton in his universal theory of gravitation. Complex questions concerning the movement of astronomical objects were resolved down to a question of masses. In calculating the Earth's movement around the Sun, nothing more than the mass of the Earth is needed; it does not matter if the Earth's composition is iron, silicon, gold, oxygen, or any other element. Nor does it matter what the Earth's shape is. Nor does it matter what the history of the Earth might be. Nor does it matter whether or not there is life on the planet. For gravitational dynamics, mass alone suffices.

The great successes of the sciences are proof of the power of isolating the physical dimensions of the world, not simply in mathematical sciences but in every natural science. In biology, questions of inheritance that had vexed thinkers since the most ancient times were decidedly resolved when attention was trained on particular chemical materials within the nucleus of the cell: DNA, the deoxyribonucleic acid. To determine, for instance, whether an organism will develop sickle cell anemia, one needs simply to analyze the strands of the genetic materials. Such success has provided humans with knowledge and powers hardly dreamed possible by previous generations. It is small wonder indeed that a mechanomorphic language with its reductionist orientation would become the dominant one throughout modern culture.

CENTURIES OF ANALYSIS have provided us with an unparalleled understanding of the role carbon plays in the ongoing wonder of life. Besides carbon, life involves primarily hydrogen, oxygen, nitrogen, sulfur, and phosphorus. To know this is to know something real and irreducible about the nature of life, something detailed, something essential.

But what about carbon? How do we learn about carbon? Carbon is not one element among others. Carbon's presence in the universe is special. This special power comes from the nature of carbon itself and is related to its intrinsic powers of being. To speak of the atomic structure of carbon, or of its various isotopes, or of its different nuclides, is to speak of the components of carbon. But carbon as a whole entity must be approached directly. To understand a thing one must understand what that thing is capable of doing in the universe. The nature of anything is shown in the role it plays in the universe. To understand something means to understand the powers it displays in the universe.

In chapter 1, "Primordial Flaring Forth," we explored the nature of a proton and the manner in which its presence cannot be simply located in a region about it. Even to speak of its most basic physical parameters requires that we include vast realms of space, even the entire universe. In a similar way, here, we are suggesting that to understand the nature of anything one can analyze it by considering that time in the past when its components were separate. One can also go forward in time and examine its future relationships. The significance or meaning of a thing depends in some sense upon the entire universe, past, present, and future.

Carbon is composed of six protons, six neutrons, and six electrons, and was assembled in the centers of stars, as we will discuss in the next chapter. All of this is essential knowledge to understand carbon. But one also needs to know that carbon possesses spectacular qualities related to thinking, to life, to aesthetic-emotional experience. Indeed, from a chemical analysis of the planet Earth, carbon's central role in life is a shock. There are great abundances of iron, nickel, and silicon; yet life uses little or none of these. There is a large amount of oxygen, so its presence in life is not as surprising. But carbon forms less than a millionth of the planet Earth. And yet, out of this, squids and anteaters and Olympic athletes come forth. Earth obviously would not have drawn carbon into its creativity were it not for the special properties of carbon.

To know carbon simply in its atomic components or in its graphite form is to have limited knowledge of the carbon element, since carbon lives and thinks in the variety of living beings of the planet Earth. We could even refer to carbon as the "thinking element" or the "element of life," where of course we are using such words as *life* and *thinking* in an analogous sense rather than a univocal sense. The thinking taking place within a human is distinct from the activity taking place within an isolated carbon atom. Yet there is some awesome relationship between the roaring activities in the carbon atom and the activities taking place in the thinking human; the storm of electricity in the carbon, that whirling presence, that incessant act makes real a power essential for the thinking process.

To understand an oak tree, we must understand that the oak tree can be analyzed down to its elements. In the same manner, to understand the elements we must learn that in their own activities are the potentialities out of which the oak tree can be evoked. Each speaks a truth about the other. The oak tree reveals the carbon; the carbon reveals the oak. So too with everything in the universe. To understand the thing we need intellectually to analyze its component parts. We also need intellectually to understand the unity of functioning of the parts within the whole. Only by confronting the thing with knowledge gotten from both directions do we arrive at full knowledge—the integral nature of the thing there before us.

The truth of the Milky Way galaxy cannot be realized by restricting our attention to its early components, hydrogen and helium. That will tell us tremendously important things about the Milky Way, but unless we also reflect on the fact that the Milky Way in its later modes of being is capable of thinking and feeling and creating, we are failing to confront the galaxy as it is. To speak of a Milky Way that does not have the inherent powers to recombine into a form capable of inner feelings is to speak of an abstract Milky Way, one that has no existence in reality. In this universe the Milky Way expresses its inner depths in Emily Dickinson's poetry, for Emily Dickinson is a dimension of the galaxy's development. If we can say with assurance that Emily Dickinson is composed of the stuff of the Milky Way, we need to say with equal assurance that Emily Dickinson in her person and in her poetry activates an inner dimension of the Milky Way.

Though we now know that at least one galaxy created a place where feelings of sublimity could be enjoyed, we will not easily learn to recognize that our feelings and our emotions pertain directly to the Milky Way. Centuries of education in an exclusive focus on the quantitative aspects of our consciousness have laid down deep habits of mind. It was such an orientation that allowed Galileo to disregard his experience of colors and of emotions and to focus his attention on the measurements of balls rolling down inclined planes. He articulated patterns of relationships between objects in the world with great detail, and in a way novel for human understanding. To ignore the psychic and sensual aspects of the universe was possible because feelings and colors were held to be manufactured by the human. What was real was the atom, its position and its velocity. Such alone were considered real because they were thought to be independent of any knowing subject.

QUANTUM PHYSICS WAS the end result of this reductionist investigation of the elementary constituents of matter; but quantum understanding was a surprise twist on the entire venture. For instance, in the understanding of classical physics, position was something that a particle possessed. The universe was a fixed container; each particle had a fixed address, its position in the world. But quantum physics implies something very different. First, there is no fixed container universe. Second, there is no address that can be assigned to each particle. Even with so simple a situation as an elementary particle we cannot speak of its position as if this were an independent abstract fact. Rather, we must speak of the position of the particle achieved by such and such an experiment. Position is an achievement of the world that involves the particle and everything that interacts with the particle. In a similar sense, we cannot speak in a scientifically viable manner about an independently existing sunrise. Sunrise is an evocation of being that involves the sun and the air and the water and sentient beings. There is no independent sunrise that a being then experiences. The universe, rather than existing in an inert objective way, is a mutually evocative reality.

Scientific knowledge in a developmental universe is no longer understood as information about an objective world out there. Scientific knowledge is essentially self-knowledge, where self is taken as referring to the complex, multiform system of the universe. The mathematically formulated designs of the scientists are not the unrestrained fantasies of humans; they refer to something ultimately real. On the other hand, the designs do not and could not exist in their mathematical formulations except for consciousness. The human is not simply noting an objective external design, but is rather intrinsically participating in the creation of these designs. That is why it is more accurate to say that the mathematical formulations of the scientists are the way in which the multiform universe deepens its self-understanding.

From the quantum perspective on the evolutionary universe, each process is ultimately indivisible. No experience can be simplistically divided up into inner and outer aspects where the outer aspects such as "position" refer simply to the objectively existing universe, and the inner refers simply to the subjectivity of the experiencing being. The elements of experience cannot be assigned a simple, univocal origin. The subjectivity of a dragonfly cannot be simplistically separated from the objectivity of its pond—the pond's very shape was decisive in shaping the dragonfly's mind. Far simpler and more persuasive is the orientation that takes the experience of the dragonfly as arising from the universe as a whole, within the unique standpoint of the dragonfly.

The sentience of an artist or a poet is not a disconnected event within a fixed container universe. In the mechanistic worldview one could believe so. The conception of a static universe with ultimately separable constituents allowed one to regard the experience of an animal as one item or fact among many others. But we now know that the interiority of any mammal, for instance, that which corresponds to parental care, is the result of a long and complex process of creativity beginning with the star-making powers of the Milky Way. Walt Whitman did not invent his sentience, nor was he wholly responsible for the form of feelings he experienced. Rather, his sentience is an intricate creation of the Milky Way, and his feelings are an evocation of being, an evocation involving thunderstorms, sunlight, grass, history, and death. Walt Whitman is a space the Milky Way fashioned to feel its own grandeur.

THE HUMAN BEING within the universe is a sounding board within a musical instrument. Our mathematics and our poetry are the merest echoes of the universe entire. We are unable to capture more than fragments, even ciphers of fragments, in our most exalted moments. Even so, as we become captivated by the quantitative aspects of our knowledge or the epistemo-logical concerns of our knowing, we often forget this deeper psychic dimension of things that activated our awareness. We enter a narrowing of human reality and take the sounding board for the whole.

Poetry and the depths of soul emerge from the human world because the inner form of the mountains and the numinous quality of the sky have activated these depths in the human. Just as with carbon, we can analyze a mountain into the form of rock and the type of mineral that compose the mountain. Mountains can also be understood as agencies in the world, participating in the ongoingness of the universe. That is, mountains act, and in a multivalent way. They sculpt the cycles of the hydrosphere and atmosphere. They shape the climates and thus the biology of the local region. And particular mountains also stun at least some of the animals. A human being, for instance, can climb a mountain and get hit by something so profound, at so deep a level, that the human will never be quite the same. This precise feeling will not occur on the ocean or in a cave or a valley. Other sorts of experience will take place there. This specific mountain moment will emerge only in the presence of the mountain; it is evoked out of potentiality by the mountain. The dynamic of the mountain is accomplishing something in the universe, is acting, is altering reality.

From our quantum perspective on evolutionary cosmology, we can approach the reality of a human stunned by a particular mountain only through a series of negatives. It is not accurate to say that the human has invented or created these feelings all by itself. It is not accurate to imagine that these feelings are present objectively in such a form within the mountain. It is not accurate to think that these same feelings would happen if a different sentient being were there, or a different mountain. The feelings are neither subjective fantasies of the animal nor simply objective experiences of the mountain. Such profound feelings, such emotions that are even tinged with personal significance and with hints of destiny, are the mutual evocation of mountain, animal, world. Depth communication of primordial existence is the reality at the foundation of all being. Humans give voice to their most exalted and terrible feelings only because they find themselves immersed in the universe filled with such awesome realities. The inner depths of each being in the universe are activated by the surrounding universe.

Without a sensitivity to primordial communication within the universe, the universe's story comes to an end. That this is certainly the case with an individual organism we can readily appreciate in the case of the monarch butterfly. Climbing out of the pupal shell, stretching its wings in the drying sunlight, what other than the voices of the universe can this butterfly rely upon for guidance? It must make a journey that will cover territory filled with both dangers and possibilities, none of which has ever been experienced before. To rely on its own personal experience or knowledge would be a disaster for the butterfly. Instead it finds itself surrounded by voices of the past, of the other insects, of the wind and the rain and the leaves of the trees.

The information of the genetic material comes forth precisely within its interactions. That is, the monarch butterfly has little if any individual awareness of the difference between beneficial winds and dangerous winds until it finds itself confronted by them in reality. The winds speak to the butterfly, the taste of the water speaks to the butterfly, the shape of the leaf speaks to the butterfly and offers a guidance that resonates with the wisdom coded into the butterfly's being. Such communication takes place beneath the level of language, even that of genetic language. It functions at the primordial reality of primal contact. The source of the guidance is both within and without—the universe as a single yet multiform mode of being.

For animals and plants, the universe is a chorus of voices. Only by heeding these can they find any chance of fulfillment in life. Only through a sensitivity to the voices of the forest can they find their way. But we are also beginning to understand in detail how the great conversation takes place within the universe community. We already treated in this chapter the way in which density waves within the galaxy initiate the birth of stars. Perhaps we can return to that context and probe the nature of listening at that level of being.

ONE NEEDS TO imagine a cloud of hydrogen and helium floating freely in a galaxy. When we say, "A density wave sweeps through the cloud," we are referring to the ripples in the space-time manifold created by the gravitational interactions as elucidated by Einstein's theory of General Relativity. But our quantitative insights into the nature of this event must not blur our awareness that we are here dealing with an ultimate act. Gravity itself is a dimension of the primal activity of the universe. Never adequately explained itself, it is a way of explaining other related activities. Gravity and its density waves, always and everywhere invisible, present themselves only in their effects. By definition this is true of all acts of ultimate mystery. The cloud of hydrogen hovers and is then transformed by something invisible that sweeps through it.

A hydrogen's manner of listening is unique to itself. Its activities are qualitatively distinct from our own. Our use of the word *listening* in this analogous sense is justified because the dynamics of an atom and of a sentient being have a formal correspondence within the understanding of quantum physics, which we now consider.

Hydrogen atoms act upon the world out of a combination of two influences. First is the influence of action elsewhere in the universe, past and present, insofar as this is immediately or mediately present to the situation. Second is that of the innate spontaneities of each particular atom, which are by their very nature beyond complete determination by the actions of others. In reference to these spontaneities physicists speak of the possibility field associated with each atom. To use Werner Heisenberg's terminology, each thing has a field of "tendencies" associated with itself, a spectrum of potentialities out of which the future will unfurl.

Though the quantum tendencies and the density waves are invisible, each acts upon the other as the wave passes through the cloud. To say that the quantum tendencies of the hydrogen are influenced by the density waves passing through the cloud, and that this shock wave initiates the cloud's implosion, is to describe an event using the univocal language of physics. An equivalently valid if metaphorical expression might be to say that the hydrogen listens to voices of the galaxy and responds by creating stars.

The advantage of the second way of speaking is its emphasis on the active role of both the hydrogen cloud and the galactic wave. The cloud of hydrogen creates the stars. For once the galaxy activates the cloud's dynamics, the cloud requires no further assistance from elsewhere; it simply responds to the astounding situation by making itself into stars. "Listening" then refers to that quantum sensitivity within the cloud that enables it to initiate an entirely new adventure in the galaxy.

43

THE CAPACITY WITHIN human awareness to hear and respond to the spontaneities of the universe was deeply appreciated by primal peoples of every continent. In order to approach their genius we need to recognize in the vast diversity of their cultural expressions an insistence on establishing a close relationship with the psychic depths of the universe. Their aim was a life in resonant participation with the rhythms of reality. For this reason the drum became their primary instrument. The drum was part of the sacred techniques for orchestrating the unity of the human/universe dance.

The drumbeat and, more broadly, the songs, chants, and dances of our ancestors expressed the visions and dreams awakened in them by the spirit world, by those dimensions of nature beyond the phenomenal world, but integral with materiality—the wild dimension of the universe. In their rituals and in their life in nature, the first peoples attended to the music sounded in their depths by the surrounding mysteries. To find oneself in the icy sea in a skin-covered kayak when the dark surface parts in a winter evening and a vast sleek presence appears, to be suddenly staring eye-to-eye with this spirit from the depths of the ocean, was cause for years of celebration. The song transmitted in that instant was the meaning and fulfillment of a lifetime.

Tonight on every continent humans will look into the edge of the Milky Way, that band of stars our ancestors compared to a road, a pathway to heaven, a flowing river of milk. Formed by the seemingly insignificant ripples in the birth of the universe, this milky band has been activating its stars with its own fluctuating waves for ten billion years, and when we stare at it, we stare back at our own generative matrix. New ripples in the fabric of space-time, we humans ponder those primal ripples that called us into being.

The vibrations and fluctuations in the universe are the music that drew forth the galaxies and stars and their powers of weaving elements into life. Not to hear such music? If autism or deafness had interrupted the music at any time in this fifteen-billion-year event, the symphony would have suddenly gone silent. Our human responsibility as one voice among so many throughout the universe is to develop our capacities to listen as incessantly as the hovering hydrogen atoms, as profoundly as our primal ancestors and their faithful descendents in today's indigenous peoples. The adventure of the universe depends upon our capacity to listen.

Humans tonight will watch the Milky Way galaxy not only with eyes, but also with radio telescopes, satellites, and computer-guided optical telescopes, with minds trained by the intricate theories of the composition, structure, and dynamic evolution of matter. Though we wait as faithfully as the ancient Inuit who stared eye-to-eye with a blue-black whale, we will not see a galactic eye blinking back at us. Though we may be as dedicated to the wild spirit of the night sky, no eye of the universe will appear from behind a cloud. Nor do we need such an experience to realize what that ancient hunter came to realize. For after such long centuries of inquiry, we find that the universe developed over fifteen billion years, and that the eye that searches the Milky Way galaxy is itself an eye shaped by the Milky Way. The mind that searches for contact with the Milky Way is the very mind of the Milky Way galaxy in search of its inner depths.

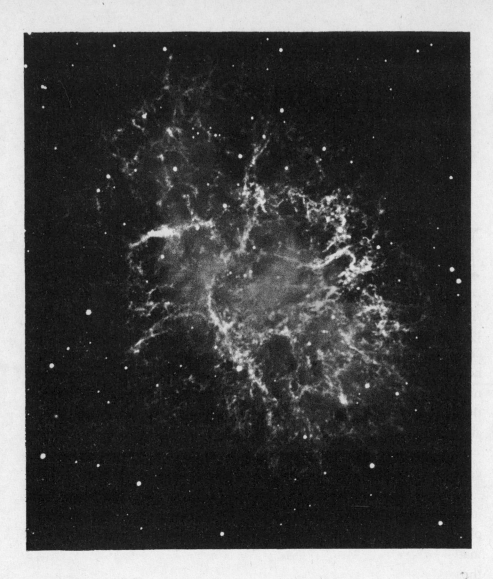

Crab Nebula, Messier 1

3.
Supernovas

IN THE BEGINNING the universe is a great shining that expands rapidly and then explodes into hundreds of billions of dark clouds. In the ensuing epoch of night, matter gathers into vast cellular regions in response to the preternatural music filling the universe. These great clouds then collapse upon themselves, each one becoming millions of times smaller. In this way, one by one, and then almost simultaneously, a hundred billion galaxies light up with a splendor new to the universe. The beginning of the universe is a smooth, intense flame. A few billion years later the large-scale structure of the universe glows in great sheets of galaxies and in their intersections in long, spidery filaments of sparkling worlds.

The story of the universe is a story of majesty and beauty as well as of violence and disruption, a drama filled with both elegance and ruin.

A billion years after the birth of the universe, when the galaxies have just emerged, great regions of hydrogen and helium drift about the center of the Milky Way. In the collapse of our galactic cloud, the spinning of the matter flattens out, disclike, as the angular rotation carries the clouds into the gentle movement of the twirling spiral galaxy. After another hundred million years the invisible density arm sweeps through the cloud and shocks it into collapsing upon itself. No further energy from the galaxy is now required. The cloud that has drifted undisturbed for eons suddenly undergoes a profound transformation that destroys its basic form but gives birth to a cluster of ten thousand diamond lights in a sea of dark night.

After its powers are activated by the density wave, the cloud reassembles its material via gravitational intensities that fracture the cloud into hundreds and thousands of self-imploding centers, the primal stars. In each of these stars, hydrogen and helium atoms are drawn together by their mutual attractions. As they collide and interact with each other, the friction creates ever higher temperatures. Their material continues to concentrate. Eventually the heat reaches temperatures beyond the endurance capacity of the atoms. Every atom is destroyed in the intensity of the implosion.

In each of the self-imploding centers the collapse continues until the pressure generated by the heat initiates the burning of nuclear fuel: millions of tons of matter are now transformed into energy each second. Each star collapses hydrogen atoms into protons, then transfigures these into helium nuclei. As the hydrogen is used up in the nuclear burning, the star turns to the helium nuclei to continue its existence. So long as a star has helium nuclei to transform into carbon nuclei, it can sustain itself against the crushing pressures of gravitation. As the helium is used up, the star turns to carbon and other higher-order nuclei to stave off its final self-implosion.

Eventually, in a million years or in several billion years, each star's resources against the collapse are all used up. If the mass of a star at this point is large enough, its gravitational pressures will destroy the star. The remaining materials will rush toward each other. Nothing in the universe can now stop them. All remaining structure is destroyed as the star implodes to a pulsar—a super-dense mass of neutrons—or collapses all the way down to a naught entity, a singularity of space and time, a black hole. This stellar being that burned brightly for billions of years, that may have showered sentient creatures with radiant energy that they transformed into their living bodies and into cathedrals that rose in wheat fields, has gone, only a black cinder left.

And yet in the great violent collapse of the star there is a surprising twist of events: the supernova. Not everything is pulled into the nothingness of the pulsar or the black hole. The neutrinos, those wispy and seemingly unimportant elementary particles, escape the collapse. As the star implodes, the neutrinos rush out in all directions to blow off the outer layers of the star, which contain the carbon and oxygen and nitrogen and other elements. Freed from the gravitational death of the star, these elements journey into the night sky, eventually to be drawn together with other elements. Under their own attraction, they form an entirely new system. A new star forms, new planets form, new life forms, and perhaps new consciousness forms; certainly a new area of the story of the universe begins out of the supernova explosion that destroys the old stellar world.

WHEN WE REFLECT on the fact that such a supernova explosion of the star Tiamat gave birth to our own existence, to our stellar system, that most of the atoms in our body were created by Tiamat and sent to us from its supernova explosion, we begin to recognize that beauty, elegance, and destruction are all layered into our ancestral origin. Our birth required drastic and vehement disruption of well-ordered communities of beings. Tiamat's blast ignited our collective adventure. Such is the nature of our story. Both infinitely patient, slow processes and sudden, cosmic intensifications are required to carry the universe through the unfurling of its material-psychic adventure.

If there is any event in the universe with the power to evoke a sense of the sublime, it is the supernova explosion. It is spectacular in the present; it shapes the future; and it has been prepared for since the beginning of time. The supernova's presence dominates at the time of its explosion, for its intensity outshines even a galaxy of two hundred billion stars. Any stellar system too close is ripped asunder. Its spreading explosion of materials governs the future, for only with such vast creativity can the adventure of the universe continue. Any part of the primordial fireball destined to become enveloped in the unfolding of life or of consciousness must first pass through this supernova explosion, an eye of the needle in the cosmic storms. And the supernova experience was dimly felt even in the first moments of the universe's existence. The primal act, constellated into the four fundamental modes of activity, established such an exquisite balance between the weak and electromagnetic interactions that the universe a billion years later would be able to express itself as supernova. Had the universe's activity ordered itself even slightly differently, the delicacy of correspondence would have been broken and the colossal act of the supernova would never have occurred.

The supernova is a multivalent event: simultaneously a profound destruction and yet an exuberant creativity. If we here attend more carefully to its violent aspects, this is only because we focus now upon the opaque or negative dimensions of the universe story. The supernova transformations rivet our attention to the negative, though of course they are only one illustration of destruction in the universe.

We have already told of the massive destruction of the early universe when all but one billionth of the world was annihilated. In the birth of galaxies some billions of years later, the density of the universe demanded collisions. Entire galactic worlds that could have conceivably given birth to seashores rich in shellfish life were blasted into shreds of gas or scattered in abandoned stars. Even galaxies in near-collision would be so violently wrenched by the gravitational tidal waves that they would remain in a disturbed condition for the next fifteen billion years.

Most of the galaxies that did survive the beginning storms were in regions so dense that their fecund spiral structures were destroyed, leaving them in the shape of elliptical galaxies no longer capable of creating stars. Such was the fate of the central galactic cluster around which the Milky Way and other local galaxies revolve. Forty million light years away, the Virgo cluster harbors more than a thousand galaxies, most of them elliptical in shape. Some of the spiral galaxies that did escape with their form intact had their interstellar gases sucked away by other worlds passing through. Had such galaxies not been crippled, they might have blossomed with living joy. Instead, the violence of the universe left them stillborn, frozen away from their possible destinies.

No galaxy, spiral or not, disturbed or not, escapes the reality of ongoing destruction. In the center of our own Milky Way galaxy a black hole churns, sucking suns into its destruction as a spider at the center of a net transforms brilliant insects into black tombs. We do not know if the suns that are swallowed by the Milky Way's black hole are living stellar systems or not, but life is only one kind of order in the universe and the destruction of such suns includes all of their orders. The creativity required to organize and sustain a star comes to nothing when a sun is flattened to a blank slate by a black hole's capture.

When we examine the nature of the universe we find galaxies with a violence so intense that we cannot account for it scientifically: great spikes of energy blasting across hundreds of millions of miles from the center of the galaxy out through its member systems. Speculation points to black holes of monstrous dimensions generating such extremes of energy from the center of the galaxy. Our own Milky Way galaxy might be capable of generating such a death spike. If such an energy blast does erupt, we will not learn of it much before it reaches us. Even now, a wall of plasma could be rifling toward us at millions of miles a minute; with such stupendous power it would not even notice the Sun and Earth as it scattered us into elementary particles in its inexorable flight of destruction through the stellar systems.

IN THIS UNIVERSE such violence! Yet when we glance at the night sky with its vast display of galaxies all swimming throughout the seeming infinity of the dark, we observe such elegance, such intricate achievement in the midst of this destruction. Is the universe ultimately destructive or creative? Violent or cooperative? The more closely we look at any place in the fifteen billion years of the universe's story, the more we realize that the universe is both violent and creative, both destructive and cooperative. The mystery is that both extremes are found together. We even find it difficult to determine when violence is simply destructive or when violence is linked to creativity.

When a female spider consumes her mate, is this a destruction on the part of the female or cooperation on the part of the male? When the mother beetle—*Micromalthus debilis*—has given birth to her offspring, the offspring begins clinging helplessly to the mother's body, but when it becomes stronger, while still in larval form, it devours the mother. What is the origin and meaning of such destruction? Such violence is not absent from the plant world either. Consider those climbing vines that begin life in utter dependency on their host tree, but which, as they grow stronger, slowly choke all life out of their host. And yet it is out of such violence— even in some mating cycles and in some processes of nurturance—that the stupendous variety displays its beauty throughout the planetary system.

VIOLENCE AND DESTRUCTION are dimensions of the universe. They are present at every level of existence: the elemental, the geological, the organic, the human. Chaos and disruption characterize every era of the universe, whether we speak of the fireball, the galactic emergence, the later generations of stars, or the planet Earth.

Wolfgang Pauli studied and gave mathematical form to a basic dynamic of the universe that now bears his name. The Pauli exclusion principle states simply that no two particles can occupy the same quantum state. This principle points out that the microcosmic world shares a dynamic observable in an everyday mesocosmic realm: matter puts up resistance. If we push two rocks together they resist becoming a single rock. Hitting a nail with a hammer drives wood apart. Two stars that collide splatter into a billion bits rather than simply passing through each other. An extreme concentration of magma blows the earth's crust into a cloud of particles.

Things in the universe put up a resistance because each being and each community of beings has its own quantum of value. Things come to be what they are out of their cosmological power of creativity. As manifestations of ultimate meaning and significance, they resist all efforts to reduce their presence in the world. Even at the level of elementary particles we find this irreducible reality of the individual. The universe, in protecting the viability of an elementary particle, works to assure the particle of its place, of its role in the unfolding story.

The resistance of matter is the first dynamic of the universe that relates to the reality of violence. The second dynamic is the need that things have for energy. The existence of any structured thing requires energy. Even something so elemental as the hydrogen atom requires energy both for its formation and for its perdurability. Scientists refer to this fact as the second law of thermodynamics, a mathematically precise way of referring to this cosmic need for energy. Any physical system closed off from new energy will inevitably decay. So for an atom, an animal, a city, an ecosystem, or a civilization to continue with its order intact requires an influx of energy in a form capable of sustaining the system.

Our universe is self-energizing. All the energy of the universe is needed by the entire universe for its own development. No development anywhere in the universe can take place without energy. The second law of thermodynamics points out that constructive activity needs energy and inevitably produces entropy, or waste. Every development has a cost, an inescapable cost. A price must be paid for creativity; an energy payment must be made for maintaining beauty; an energy payment is demanded for any and all advance. This is the second fact concerning the structure of the universe that we must ponder if we are to understand the nature of violence.

In addition to the resistance in the universe and the cost of creating order, there is, thirdly, the tendency in all things toward fulfillment of their inner nature. In physics this is referred to as the quantum tendencies that hover within any physical situation. In cybernetics this is referred to as the autopoiesis of a coherent system, such as a developing star or a mature ecosystem. In biology, this is referred to as the epigenetic pathways folded into a particular ontogeny. Each acorn has layered into it the future possible destinies of the oak tree. When the branches are not getting the sunlight so deeply desired they will spread out wider, they will reach up higher in a struggle to find what they need. Unless these tendencies are taken into account there is no way to understand an unfolding organism as it interacts with its environment and its own genetic inheritance. All things in the universe, in their own subjectivity, are pervaded by inherent tendencies toward fulfillment of their potential.

RESISTANCE, ENERGY, DREAMS: these are the sources of all violence. Another way of describing them is to speak of past, present, and future. Opacity or resistance comes from an insistence that accomplishments of the past be preserved against attempts to remove them. Energy—the cost of creativity—points to the finite nature of the present universe. Since energy is required to sustain anything whatsoever, a decision must be made concerning what we energize in the present moment. What we choose to energize will persist and what we choose not to energize will perish. Dreams refer to the unborn, to the darkly felt inclinations toward a new world, a not-yet world. The future as not-yet works in the present by making a bid for a quantum of energy necessary for its fresh and novel embodiment.

The three terms *resistance, energy,* and *dreams* have been explored with great care within our scientific enterprise. As mentioned above, the resistance of the universe, in the language of physics, is called the Pauli exclusion principle. The cost of systemic perdurability is formally expressed in the second law of thermodynamics. What we are calling dreams is referred to with the phrase *quantum tendencies* in the theory of quantum mechanics.

Our empirical investigations of the structures of reality reveal a universe where opacity, energy-need, and organizing tendencies are fundamental. These three imply that violence and destruction are also fundamental to reality.

The desires of a single pair of aphids, if satisfied, would destroy the Earth. After only a single year such aphids would generate over half a trillion offspring. The same could be said for the majority of the insects and, in a related sense, much of life. Their desires to unfold into being are immense, while the energy of the Earth is finite. To energize those aphids means to remove energy from other regions, from other beings, and for these others to perish. Each being in the universe yearns for the free energy necessary for survival and development. Each existence resists extinction. The consequent history of violence in the universe is as inevitable as the gravitational pull between the Earth and the Sun.

THE UNIVERSE THRIVES on the edge of a knife. If it increased its strength of expansion it would blow up; if it decreased its strength of expansion it would collapse. By holding itself on the edge it enables a great beauty to unfold. The Milky Way also thrives at the edge of a knife. Decrease its gravitational bonding and all the stars scatter; increase the gravitational bonding and the galaxy collapses upon itself. By holding itself in the peace of a fecund balance of tensions, it enables planetary structures and living beings to blossom forth.

Every being that thrives does so in a balance of creative tension. The boreal forest that receives too little or too much sunlight will collapse into a different biome. The universe is a multilevel web of different communities of beings with ancient and well-established orders of bonding. This beauty evaporates when the enabling relationships are too severely disrupted.

Yet the universe brings forth new centers of creativity into this world of established relationships and long-honored traditions. Potentially infinite desire finds itself within a woven fabric of finite energy. This condition holds at every level of reality. The mollusk in the sand of the ocean, the bacteria in the rotting redwood tree in the forest, the tornado in the wind currents of the summer drought, the black hole in the center of the galaxy—each exists with demands in a world tight with constraints on the very energies necessary to satisfy these demands.

Working within these structured communities and their many imposed limitations, the universe brings forth both its violence and its creativity. These obstacles, these boundaries, and these limitations are essential for the journey of the universe itself. The clouds of hydrogen discover themselves in a field of mutual attraction that leads to an ever-increasing temperature due to the natural opacity of hydrogen. But out of this obstacle to their movement, the star is created. The existence of the star arises out of the constraints placed upon the imploding particles. If they could continue without ever meeting resistance, no star would ever emerge. Mountains, too, result from such opacity. The tectonic plates on the surface of the earth crash and crumple into each other. Could such resistance be removed, the continents would never buckle or fold over into the majesty we call the Alps or Himalaya or Sierra.

In the primeval fireball, which quickly billowed in every direction, we see a metaphor for the infinite striving of the sentient being. An unbridled playing out of this cosmic tendency would lead to ultimate dispersion. But the fireball discovered a basic obstacle to its movements, the gravitational attraction. Only because expansion met the obstacle of gravitation did the galaxies come forth. In a similar way the wings of the birds and the musculature of the elephants arose out of the careful embrace of the negative or obstructing aspects of the gravitational attraction. Any life forms that might awake in a world without gravity's hindrances to motion would be incapable of inventing the anatomy of the cheetah.

THE OBSTACLES TO a microphase desire can be understood as the presence of the macrophase structure. By microphase we mean that which pertains to the here and now of a particular creature. By macrophase we point to the larger realities involved in the moment, both in terms of the largeness of the universe and of Earth and the mystery of the unborn future. The obstacle that one hydrogen atom presents to another is required if a star is to emerge in the future. Of course, from the perspective of a cloud of hydrogen atoms, the idea of a future star is absent. Nevertheless we can, after the fact, ponder these initial resistances and understand them as the first functioning of the future structure of the star. In a similar way, the spectacular structured sensitivity of the eye of a deep-sea fish can be understood as the meaning of the difficulties that were experienced by the earliest sea worms searching for sustenance in the dark ocean bottom.

If the universe is to have the macrophase structure of photosynthesis, or if Earth is to bring forth the osmotic powers of the mushroom, then there must be earlier beings who ram up against obstacles to their microphase desires. Many of the inventions of the natural world arose out of beings meeting the constraints of the universe with creative responses. Only by dealing with the difficulty does the creativity come forth. The violence associated with the hawk starving to death or the vole being consumed are intrinsically tied to the creativity of each. The beauty of their response arises from an inherently difficult situation.

WHILE WE MIGHT go on with further explorations of violence and creativity within the universe as a whole, we need to pay particular attention to violence and creativity within the human world. For with the human a new order of being arises. Indeed every invention of the universe having macrophase significance transforms the nature of violence and creativity. Violence, destruction, and disruption, on the one hand; creativity, synthesis, integration on the other; these are multivalent words having distinct meanings for the galactic period, the stellar worlds, the elemental realm, the organic and the human realms.

Within self-reflexive consciousness, terror becomes aware of itself. With such conscious self-awareness, life understands that it is precious and liable to destruction. It is out of this new depth of fear that humans devoted themselves to eliminating violence and destruction. A new kind of insecurity emerged in the universe with humans. The elegance of the balanced world burst into awareness of itself, and the accompanying terror at times proved to be too much for humans to manage creatively.

The determination to dominate the universe so that all insecurity, limitation, destruction, and threat of destruction could be eliminated eventuated in racism, militarism, sexism, and anthropocentrism, dysfunctional maneuvers of the human species in its quest to deal with what it regarded as the unacceptable aspects of the universe.

The question is whether the cosmos requires re-engineering in its basic form, so that whatever we find undesirable must be altered. Now on one level, of course, we are going to pursue that which we find desirable. But our special difficulty shows itself when we fail to take the macrophase meaning of the Earth and universe into our microphase species projects. If humans were a relatively minor crustacean species, perhaps considerations of the macrophase dynamics of the universe would not have to become a central focus of consciousness. But our decisions have immediate effects on the macrocosmic reality throughout the Earth community. We ponder the cosmological significance of our coherence within the larger community because the enhancement of the Earth community as a whole seems to require this of us.

Resistance, energy, and dreams fill the universe and Earth in a multiform manner. Recent humanity has attempted to relate to these cosmological realities by destroying resistance, denying intrinsic cost, and magnifying the intensity of all its desires. If we meet resistance, we work to eliminate it. If we are asked by the universe to make the payment necessary for development, our response is to avoid the bill. On the other hand, if we discover any new human desires, we invest tremendous effort in fanning these desires, no matter how superficial and costly they are for other members of Earth. Our refusal to accept any limitations for movement in space has resulted in our destructive transportation systems, imposed on the planet without regard for the rest of the Earth community. Our refusal to accept any limitations on our desire for consumer items has resulted in the disruption of ecological communities throughout the planet. Our desire to have children without regard for the capacities of the biome has resulted in the explosion of human numbers and the consequent sufferings inflicted upon so many billions of them.

The obstacles—both physical and psychic—that contemporary humans experience can indeed be demolished or denied for another generation, enabling us to continue with our destruction. Some of this will certainly take place as those on the hunt for material gains exhaust their inventiveness searching for ways to pretend the universe can be fooled. But another orientation is possible. We might come to see in the very constraints themselves the presence of a future sublime beauty, struggling to emerge. In a later chapter we articulate some of the contours of this beckoning dream, the Ecozoic era of the Earth community. For here we can comment on the relationship of the Ecozoic era to our present realities.

If the atoms in the prestellar cloud had been given language and the power to reflect upon inner experience, so that they could ponder the significance of the density waves sweeping through them and the rush of atoms ramming up against them, they would even then not have been able to speak in clear terms about the star they were destined to become. Even if they had been able to speak with intelligence about the new tendencies that gripped them, they would never have been able to point in any direct way to the great supernova burst whose existence they would enable. The future is undetermined, always a surprise.

We have at best only a vague and dark sense of what draws out our creative energies. The beauty of the star gripped the atoms in some primordial manner; the beauty of the new flowering of Earth's realities likewise grips us and is in many ways the central significance of all our experiences of obstacle, disappointment, dismay, and despair. We do not know and cannot know the exact shape and form of the Ecozoic era. We cannot know with certainty even what is required of us now.

And besides the very unborn nature of the future, we must also deal with the inadequacies of our present forms of consciousness. For even if the future were determined, its complexity would be beyond our capacity to articulate it. We will find our way only with a deep and prolonged process of groping—considering with care a great variety of interpretations, weighing evidence from a spectrum of perspectives, attending with great patience to the inchoate, barely discernible glimmers that visit us in our more contemplative moments. Out of this welter will slowly emerge our way to the star.

P UTTING TO ONE SIDE the detailed action-plans future humans will eventually settle upon, we propose here a radically different orientation within the dynamics of the universe. We can begin to understand that in the very resistance to our former habits lie the central directions for our creativity. By embracing these obstacles and the suffering these sometimes entail, we can participate more effectively in giving shape to new forms and patterns of human existence.

Can modern humans sacrifice some of their ephemeral pleasures for the well-being of the Earth community as a whole? To do so requires our reaching an insight found in a pervasive manner throughout all the earlier forms of human understanding—that the universe has what can be called a sacrificial dimension. When we reflect upon the omnipresence of destruction and violence throughout the layered universe, and on the mysterious relationship of this destruction to the evocation of a great beauty, we can begin to approach such an understanding. Life includes, in its essence, hardships of many kinds. To refuse these, to refuse to accept what might be called legitimate suffering, is to opt for a reduced existence.

An individual who takes as a central life project the avoidance of pain and suffering of any sort will lead a neurotic and ephemeral life. A society that takes the elimination of all hardship and all suffering as the essential aim of its major social institutions will create a flat existence for its humans and a deleterious world for its nonhuman surroundings. And the total cost to be paid in either case is monstrous, enormous, grotesque.

The word *sacrifice* may be too encrusted with a history of misuse to serve us in our struggle to understand what is required for us. But for our ancestors a sacrificial act was a way of making holy, especially when the act had a bitter dimension to it. That such an insight is not so very distant from modern humanity is evidenced in our popular culture generally. The hero in our literatures who takes on deprivations, who sacrifices comfort, who throws off status or wealth, all in an effort to secure the safety and well-being of others, is not far in nature from the shaman who sacrifices something valuable, even his or her own physical well-being, as a way to assist in the empowerment of others. And our own immediate and high esteem for heroic types who rush into flaming homes or dive into freezing rivers to rescue someone, who sacrifice wealth or prestigious careers to work for the betterment of the Earth community, reveals our own recognition that the individuals who act this way make clear a sacred dimension of existence.

The pervasive presence in the human world of both sacrifice and the honoring of those who sacrifice seems to indicate our awareness that suffering and destruction are intimately associated with existence itself. We are moved to take on health-restoring deprivations because of our recognition that existence is a high privilege. Even in ritualistic sacrificial offerings, humans enter in a formal way the ongoing sacrificial reality of the universe. In all this the human community is attempting to give recognition to a central dimension of existence and to enter this reality in a creative rather than an unconscious and destructive manner.

The primal human insistence upon sacrifice can be understood as an early intuitive grasp of the essential truth in the second law of thermodynamics. Rather than speaking of the movement toward entropy, the primal peoples would speak of the intrinsic pain that accompanies so many genuine advances. The pursuit of pain for its own sake is as pathological as a life dedicated to avoiding pain of any sort. The mystery of existence at times asks terrible things of us, suffering beyond understanding. The hope in such bitterness is the creative resolution somehow intrinsic to the suffering.

In no sense do we mean to suggest that the cruelties taking place in the Earth community today will be justified by the emergence of what is to come. No intellectual resolution of the conundrum of suffering is possible, or even desirable. The tension of existence in time within the phenomenal world is a primordial aspect of our existence. The bafflement of suffering and violence is part of the resistance, part of the obstacle that we discover in our lives. Great art, monumental speculative philosophy, profound institutional and social reform, epochal works of music, and world-transforming technical inventions have been created by individuals stunned by suffering and violence. To eliminate the tension would be to eliminate the beauty.

I N THE UNBEARABLE pressures of a star, hydrogen is burned into helium, helium is burned into carbon, carbon is burned into oxygen. Anything available as fuel is shoveled into the nuclear blast furnace to stave off gravitational implosion. But after billions of years of striving in this way, Tiamat found herself pressed to the wall, exhausted by the effort, helpless to do anything more to balance the titanic powers in which she had found her way.

When her core had been transformed into iron, she sighed a last time as collapse became inevitable. In a cosmological twinkling, her gravitational potential energy was transformed into a searing explosion, a single week-long flash of brilliance that would catch the attention of every watchful creature in the galaxy. But when the brilliance was over, when Tiamat's journey was finished, the deeper meaning of her existence was just beginning to show through.

Out of the spectacular tensions in the stellar core, Tiamat had forged tungsten, copper, and vanadium. She vanished as a star in her grand finale of beauty, but the essence of her creativity went forth in wave after wave of fluorine, astatine, and bromine. Tossed into the night sky with the most extravagant gesture of generosity were cesium, silver, and silicon. Tiamat had evoked magnesium, osmium, gallium, rhodium, and titanium—each a new world of power cast forth by the quintillions for the future unfolding of the universe. For any worlds intelligent enough to receive them were oceans of palladium, germanium, and cadmium. None of these power elements had appeared in the primeval fireball or in the early galactic era. These beings were new, and though they mixed inconspicuously into the great dark clouds that wafted out from Tiamat's explosion, in their essences, in their potential power, they glittered as brilliantly as the supernova incandescence.

Tiamat had forged calcium, a new presence that would one day support both mastodons and hummingbirds. Tiamat had forged phosphorus, which would one day enable the majestic intelligence of photosynthesis to appear. Tiamat had sculpted oxygen and sulphur, which would one day somersault with joy over the beauty of Earth. Great destruction, unbearable violence, and out of this Tiamat invented the cosmic novelties of carbon and nitrogen, two astounding powers that would one day sparkle as life, as consciousness, as memories of beauty laced into the genetic codings. Tiamat's story—the story of her brilliance, her creativity, her passion, her destruction—is a sacred intensification of the universe's journey. In her story we witness a burst of glory, an amplification of the universe's beauty, and a dangerous and joyful release of power.

Tropical Storm Sun, Indian Ocean

4.
Sun

WHEN THE UNIVERSE flared forth it was capable of a particular spectrum of activities. When the laws for the fundamental particles were stabilized, the universe's capability for creative action advanced to a new set of rich possibilities. With emergence of the primordial atoms and the galaxies the creative capabilities were again transformed to a new level. The hydrogen and helium and galactic structures altered the large-scale dynamics as well as the small-scale nature of the universe. With the beginning of each new era of the universe, activity and its multiform possibilities undergo a creative transformation.

Galaxies after supernovas are qualitatively different from galaxies before supernovas. When the stars implode and the neutrinos blast out, vast remnants of the stars are sent streaming away at such high velocities that in a relatively short time these new elements are thoroughly mixed into the swirling gases and stars of the galaxy. Stars of second, third, and higher generations arise out of a fundamentally different galactic matter.

In the Milky Way galaxy four and a half billion years ago, the disc of stars was relatively rich in all the elements. Rushing through this sea were the two great spokes of the density arms whose velocity at the edge of the Milky Way was twenty miles per second. As these invisible waves spun, they drew forth millions of star bursts, each new star with its own particular destiny. But now we focus on one branch of this galactic evolution, the one leading to our Sun.

In the area of space-time where the Sun emerged, the density wave passed through every one hundred million years. The most massive stars exploded as supernovas and enriched the interstellar matter. By the time of the Sun's birth, perhaps one hundred passes of the star-making wave had occurred. Eventually a wave swept through that triggered the burst of some ten thousand stars all at once. This cluster of stars ranged from blue giants with surface temperatures in the range of fifty thousand degrees, able to burn so brightly but exploding after only a million years, all the way down to small cool stars with surface temperatures of only a couple of thousand degrees, but capable of burning for hundreds of billions of years. Ten thousands stars, and each had a different fate determined by the waves fracturing the initial cloud. One of these subclouds was destined to become the Sun of our solar system, all its trillion pieces without any awareness of the amazing journey that had been chosen for them.

In the beginning, the solar cloud was distinguished from other sub-clouds primarily by size. Larger clouds were fated to become blue giants that would quickly move into collapse or a red giant phase; the smaller clouds were headed toward long and relatively cool existences as white and brown dwarf stars. But the solar cloud, five million times the diameter of the Sun that this cloud would eventually spawn, headed toward an adven-ture near the center of the main sequence of stars.

This formation would be finished in five hundred thousand years. The subcloud, freed from interactions with the others, gathered itself into a state of increasing nonequilibrium; matter fell into a center, creating a great deal of heat that radiated out from the collisions. After several hundred thousand years, the central core became thick enough to trap its own light and began to heat up even more rapidly. The core was in size the same as Jupiter's orbit around the Sun today, but comprised less than one percent of the total mass of the solar cloud.

When the core's temperature reached two thousand degrees, the hydro-gen molecules melted down into single atoms and the core collapsed in upon itself still further, generating more heat. As the matter continued to collapse down and pound on this inner core, the temperatures rose until all atoms were ionized back into their elementary particles, which brought about the final collapse. The pressure of the plasma core grew to match the gravitational pressure. When the temperatures reached ten million de-grees, the hydrogen at the center began to form helium. Our Sun was born.

In a single cloud, ten thousand simultaneous births would include a certain amount of disruption between the implosions. Protostars colliding with other protostars would be sufficiently disturbed so that some of the star formation would never complete its work. In time most of these sub-clouds drifted apart.

The vast majority of the gas that did not make its way to the Sun's core would be blasted away. In addition to the intense radiation of its light, the Sun also sends out each day twenty billion tons of protons in the form of solar winds. Though obscured in its first half million years, the Sun cleared much of the remaining gas out of its way. Most of the gas was blown off, but not all. Spinning around the Sun was a disc of the original subcloud just large enough to resist these cosmic rays from the Sun. Only one one-hundredth the size of the Sun—a cool remnant of the subcloud, a han-ger-on, a residue, something left over when the main action of the star formation was finished—this swirling disc of elements gave birth in time to Mercury, Venus, Earth, Mars, Jupiter, Saturn, Uranus, Neptune, and Pluto.

Even to list these names evokes another dimension to our discussion. The names remind us of the history of attempts to orient humans in the universe. Until the last sentence of the above paragraph, we were in the world of protons and stellar collapse, the world of quantities—six thou-sand degrees, the speed of light, 1 percent of the mass—with its connota-tions of impersonal forces, probabilistic outcomes. But to evoke the name of a god—Jupiter—or of a goddess—Venus—reminds us that in speaking of the birth of the Sun, or of Jupiter, or of Earth, we are dealing ultimately not just with the self-assembly of a ball of gas but with the deep mysteries of existence itself.

The existence of the Sun and of the planet Earth beckons us to tell their story. Most peoples have their own story of the Sun, its birth, and its sig-nificance. We come after so many of these cultures and their stories, just as the Sun came after so many supernova explosions. A new moment for telling the Sun's story is possible. All the ancient questions we ask again. What is the meaning of the Sun? How do we relate to the Sun? As a deity? As a ball of thermonuclear reactions? Is the Sun special in any way? Does it deserve its name, as former peoples designated it *Apollo,* or *Sol,* indicat-ing a solar personality? Or is it rather that the sun is identical to billions of other stars; that its development matches their development? If indeed the Sun's development is somehow different, how is it different and how is it similar?

Our universe story now concentrates on the details of one particular strand. To answer the above questions means to articulate our assumptions concerning the evolution taking place elsewhere, outside our detailed understanding.

I N OUR WESTERN scientific tradition, beginning assumptions are framed in terms of overriding principles. Such interpretations of reality serve as starting points to further investigation. Based on available evidence, they appear to us as reasonable, even though they are not proven empirically. The foundational principle in cosmology was given its definition in 1931 by Albert Einstein. Called the *Cosmological Principle,* essential for the entire research enterprise in cosmology around the planet today, it states simply that "All places are alike." This is a *principle* rather than a *fact* because we have knowledge of it only from one perspective, that of our Earth-based observations. From our viewpoint the universe seems to be the same or nearly the same in all directions. From every direction, microwave background radiation left over from the fireball comes to us in nearly identical packages. Also, in every direction we find a sea of galaxies. The occurrence of galaxies is irregular on small scales but homogeneous around distances of a billion light years. On such scales the universe looks the same in all directions.

We have no actual experience of what the universe looks like from the Hercules cluster. We assume it is the same and refer to the Cosmological Principle as validating our assumption.

The Cosmological Principle is spatially oriented—every point in space is the same as every other point. What we want to establish here is an extension of the Cosmological Principle. From Newton up until Einstein, mathematical cosmologists worked in ignorance of the evolution of the universe as a whole. The basic concepts were thus forged in the context of a cosmos that was unchanging as a whole. We now know that we live in a cosmogenesis, an ongoing, developing reality. Our extension of the Cosmological Principle, which we will call for clarity the *Cosmogenetic Principle,* assumes that every point in the universe is the same as every other point, and additionally, that the dynamics of evolution are the same at every point in the universe. The Cosmological Principle holds that the distribution of matter and energy is basically the same throughout the universe; the Cosmogenetic Principle further assumes that the dynamics of development are basically the same throughout the universe. As with the Cosmological Principle, the Cosmogenetic Principle is not something to be proven. It is a fundamental assumption based on the evidence that we have of the universe's development. Only someone who has visited every place in the universe could say from immediate observation whether or not such principles are true. Based on our own observations, however, the Cosmogenetic Principle appears to be a reasonable assumption.

Cosmogenesis, as well as its microphase complement, epigenesis, refers to structures evolving in time. Cosmogenesis pertains generally to large-scale structures such as galaxies and stellar systems, while epigenesis refers to the development of forms within the life world. What we observe is that forms and structures in the universe arise, evolve in interactions, achieve stable if nonequilibrium processes, and then decay and disintegrate. The Cosmogenetic Principle simply states that the dynamics involved in building the structures that appear in our own region of space-time permeate the universe as well. This principle does not hold that the actual structures in one locale are identical to the structures elsewhere in the universe. The Cosmogenetic Principle simply states that the form-producing dynamics at work here are also at work, or are at least latent, everywhere else in the universe. It states that our region is the same as every other region in the universe insofar as these dynamics are concerned.

Our principal paradigm of form production is the Sun. A cloud of nearly homogeneous composition drifting along in a near-equilibrium state was shaped into the Sun, a highly nonequilibrium, evolving structure. Such stellar structures pass through any one of a spectrum of sequential stages, the particular sequence depending upon the initial state as well as the conditions of the star's environment. After existing for a period, the star loses its form altogether, collapsing into a burnt-out dwarf, or disintegrating into either a pulsar or a black hole. As far as we can observe, the powers that activate a star's existence function throughout the spiral galaxies of the universe.

In some galaxies, as we have mentioned, the discs have been swept clean of interstellar gas, and in some other galaxies the spiral structure has been destroyed; in either case these galaxies only rarely create new stars.

The Cosmogenetic Principle implies that had such galaxies not been distorted, they would have possessed the ability to generate stars. The powers for assembling stars are assumed to be present, at least latently, in the sense that if the spiral structure were somehow restored and interstellar gas supplied, stars would soon sprout forth.

Another central example of cosmogenesis is the birth and development of galaxies. As far as we can determine, most of the galactic structures were created throughout the universe some billion years after the primal Flaring Forth. We assume that the power to generate galactic structure is everywhere present in the universe. Still another example of form-production in the universe is the emergence of the primordial atoms. The Cosmogenetic Principle assumes that the structure of the hydrogen atoms on Earth is similar to, even exactly the same as, the structure of hydrogen atoms throughout the rest of the universe. There is no way we can be certain that this is true without actually going elsewhere and inspecting hydrogen, but there is no evidence in favor of the proposition that hydrogen is different elsewhere, and it violates our intuitive sense that our location in space-time should be similar—in its basic interactions and dynamics—to places throughout the universe.

We can further elucidate the meaning of the Cosmogenetic Principle by comparing it with another important principle, the second law of thermodynamics. The second law refers to the tendency in any closed system to increase its entropy. For instance, the behavior of heat is not reversible in time. According to the second law, a concentration of heat will disperse in all directions and, once dispersed, the heat will never gather into a concentrated state.

The Cosmogenetic Principle is complementary to the second law. The second law refers to the dynamic in the universe that breaks down order; the Cosmogenetic Principle refers to the dynamic of building up order. We can see these two principles best illustrated by the story of the Sun.

We began with a stream of hot atoms rushing through space from a supernova explosion. In each instant they radiated photons, which reduced their temperature. As they collided with cooler atoms they gave off even more of their energy. In time, they mixed into the other atoms to become a homogeneous cool cloud near equilibrium; no part of the cloud was significantly hotter than any other part. This entire movement from the hot stream of atoms to a quiescent equilibrium state is an illustration of the second law.

If such a cloud is large enough, and within the disc of a spiral galaxy, it can be shocked just slightly to activate the star-making dynamics of cosmogenesis. Thus did a reverse sequence take place. Rather than a situation illustrating the second law's movement to an equilibrium state, we found the atoms moving to a highly nonequilibrium condition—an extremely hot region in one place that persisted throughout billions of years. The Sun's powers hold it in this nonequilibrium state. If at the end the Sun were to explode, and were to send streams of hot atoms out in all directions, we would be back to the beginning of our scenario.

BEFORE SCIENTISTS DISCOVERED cosmogenesis, it was only natural that the second law of thermodynamics would be seen as the ultimate principle. No one was yet aware of large-scale form-producing powers that would over fifteen billion years give rise to every structure in existence. Perhaps now that we are studying the ongoing development of the universe, it will be the Cosmogenetic Principle that will be seen as the ultimate. Perhaps in the twenty-first century, scientists will become so fascinated with the order-generating powers of the universe that they will tend to forget the nineteenth-century's focus on the second law. In fact, either principle taken by itself is only a partial insight into the nature of the universe. To ignore either the second law or the Cosmogenetic Principle is to ignore the way the universe actually works.

BECAUSE FOR CENTURIES we have focused attention on the form-destroying powers of the universe, and are only now beginning to investigate form-producing dynamics, our descriptions will be speculative, open to further development as our understanding grows. Our approach will be based primarily on observation. We find ourselves in a universe that has over fifteen billion years given birth to an astounding development. What can we say about this?

Most central of all, perhaps, is our knowledge that in some sense the structures of the universe were "aimed at." The structures are not entirely accidental in the sense of being the result of random collisions in an otherwise indifferent universe. To get atoms in the universe to bounce together haphazardly to form a single molecule of an amino acid would require more time than has existed since the beginning, even a hundred times more than fifteen billion years. Yet amino acids formed not only on the planet Earth, but throughout the Milky Way galaxy.

The Cosmogenetic Principle assumes that the self-organizing dynamics that fashioned the amino acids are of course not special to the Milky Way but are present in every galaxy. Once such powers are activated, they set to work with great efficiency. The powers involved in making an amino acid involve electromagnetic, gravitational, and strong and weak nuclear interactions. The algorithm for weaving these interactions into a process producing amino acids is what we are referring to as a *form-producing power*.

We have chosen the amino acid to illustrate the way the universe aims at particular structures only because it is a famous example in scientific literature. In fact, if we consider any of the structures of the large-scale universe, we discover that they are difficult or impossible to account for in a random or unbiased cosmos. Certainly this is true for the galaxies. If we consider an indifferent universe in either a chaotic or an equilibrium state, the chances that a galactic structure will evolve within a billion years are negligible. With one hundred billion years the chances are still negligible. The obvious conclusion, enshrined as the Cosmogenetic Principle, is that in this universe there are entirely natural powers of form production that, when given the proper conditions, will create galaxies.

THE FACT THAT form-producing powers are latent everywhere in the universe is the first feature of the Cosmogenetic Principle. The second is the relationship between such powers through time. Just as we need a cloud of atoms at a particular density before star-making can begin, so too, we need a particular sequence of formations before the next level of activity can begin. For instance, it is impossible to activate star formation in the primeval fireball. Only a coordinated sequence of transitions makes possible the emergence of entirely new realities.

Such a sequence of transformations in the evolution of a star is a series of decisions on a branching tree of possibilities. The first breakpoint is the arrival of the shock wave. The nature of the cloud and the nature of the wave will determine which of several new branches is taken—one leading to a yellow star, a giant blue star, a small red star, or no star at all. The next branchpoint comes when hydrogen is done burning. Among several paths, one must be chosen. Perhaps the powers of burning helium will be activated. Perhaps the rapid transformation of a helium flash will reduce the star to a white dwarf. Or perhaps something from the environment will alter destiny by adding or subtracting a large amount of mass.

The movement into the future through one sequence of these pathways is free, in the sense that at each branchpoint in the universe a fundamental decision is made determining that particular direction. But the movement is also determined, in the sense that at each branchpoint only a particular set of options presents itself. The Cosmogenetic Principle states that vast webs of pathways exist potentially at every place in the universe; the precise segment of these possibilities to be explored will depend upon the history of the region's creative development. We cannot know what specific features have come forth in any particular region unless we go there and observe. But since we now possess extensive data on the development taking place over fifteen billion years, we can begin to articulate some general features of cosmic evolution. These features, we expect, are generic characterizations of the unfolding taking place anywhere in the universe.

THE COSMOGENETIC PRINCIPLE states that the evolution of the universe will be characterized by *differentiation, autopoiesis,* and *communion* throughout time and space and at every level of reality. These three terms—differentiation, autopoiesis, and communion—refer to the governing themes and the basal intentionality of all existence, and thus are beyond any simple one-line univocal definition. The next sections articulate some of the meanings we associate with the terms and might be considered a prologue for later treatments as our direct experience of the universe's development extends throughout space and time. Some synonyms for differentiation are diversity, complexity, variation, disparity, multiform nature, heterogeneity, articulation. Different words that point to the second feature are autopoiesis, subjectivity, self-manifestation, sentience, self-organization, dynamic centers of experience, presence, identity, inner principle of being, voice, interiority. And for the third feature, communion, interrelatedness, interdependence, kinship, mutuality, internal relatedness, reciprocity, complementarity, interconnectivity, and affiliation all point to the same dynamic of cosmic evolution.

These three features are not "logical" or "axiomatic" in that they are not deductions within some larger theoretical framework. They come from a post hoc evaluation of cosmic evolution; these three will undoubtedly be deepened and altered in the next era as future experience expands our present understanding.

The sequence of events in the universe becomes a story precisely because these events are themselves shaped by these central ordering tendencies—complexity, autopoiesis, and communion. These are the cosmological orderings of the creative display of energy everywhere and at any time throughout the history of the universe.

The metaphor of music assists in expressing the nature of this ordering. From one perspective, a symphony is a series of notes and silences, a sequence of disturbances in the air, a string of tones occurring in a certain interval of time. So too, from one perspective the universe is a series of occurrences, a sequence of disturbances in the field of energy throughout reality, a string of material and energetic configurations taking place in an interval of time.

From a deeper understanding, the notes are ordered as they are to give fresh expression to the underlying themes of the symphony. The notes occur in the way they do so that something that is otherwise silent and ineffable can be sung into existence. Music consists of both the particular notes and the governing themes. For without the notes the themes would have no power to move anyone; but without the themes the notes would only irritate and distract.

The universe arises into being as spontaneities governed by the primordial orderings of diversity, self-manifestation, and mutuality. These orderings are real in that they are efficacious in shaping the occurrences of events and thereby establishing the overriding meaning of the universe. Indeed the very existence of the universe rests on the power of this ordering. Were there no differentiation, the universe would collapse into a homogeneous smudge; were there no subjectivity, the universe would collapse into inert, dead extension; were there no communion, the universe would collapse into isolated singularities of being.

COSMOGENESIS IS ORDERED by *differentiation*. From the articulated energy constellations we call the elementary particles and atomic beings, through the radiant structures of the animate world, to the complexities of the galaxy with its planetary systems, we find a universe of unending diversity. Where we once thought of the heavens as filled with stars, we later identified planets, the wandering stars, as differing from all other stars; then we noticed the nebulae and galaxies that were different still; then we noticed how each planet was different from all the others. The more intimately we become acquainted with anything, the clearer our recognition of its differences from everything else. Certainly our awareness of the differentiation of the universe must be regarded as a central achievement of the human adventure.

There has never been a time when the universe did not seek further differentiation. In the beginning all the particles interacted with each other with minimal distinction. But with the cosmic symmetry breaking, four branches of interactions differentiated from each other. From the thermal equilibrium of the fireball the universe sprouted into a universe differentiated by galaxies, with no two galaxies identical.

Not only is each thing new and different from all other structures; the dynamics of the new structures are qualitatively new as well. The interactions governing the elementary particles are qualitatively and quantitatively distinct from those involving atoms, which are again different from the dynamics involved with stars and galaxies. There are also similarities between levels. But the point to be emphasized here is that the generic form of the equations on one level is particular to that level. Each level of the universe is a distinct "world."

Differentiation, autopoiesis, and communion refer to the nature of the universe in its reality and in its value. These three features are themselves features of each other. So, for instance, *differentiation* also refers to the quality of relations taking place in the universe as they differentiate after

73

the symmetry breaking. This diversity of relatedness pertains to human knowing as well—knowledge represents a particular relationship we establish in the world. For knowledge or understanding to be reduced to one-dimensionality—as with certain scientist tendencies to reduce all knowledge to its quantitative mode—would be similar to reducing a full symphony down to a single note. An integral relationship with the universe's differentiated energy constellations requires a multivalent understanding that includes a full spectrum of modes of knowing.

The multiform relatedness demanded by a differentiated universe rests upon the fact that each individual thing in the universe is ineffable. Scientific knowledge refers ultimately to the way in which structures are similar, whether of stars, or atoms, or cells, or societies. But in the universe, to be is to be different. To be is to be a unique manifestation of existence. The more thoroughly we investigate any one thing in the universe—the Milky Way galaxy, the fall of Rome, the species on a particular tree in a rainforest—the more we discover its uniqueness. Science simultaneously deepens our understanding of a thing's structure and its ineffable uniqueness. Ultimately each thing remains as baffling as ever, no matter how profound our understanding.

The universe began in a concentration of energy and at each instant has re-created itself new. The seemingly infinite power for transfiguration in every region of the universe speaks of an inexhaustible fecundity at the root of reality. When we examine the entire display we find, pervasive with being, an insistence to create anew.

What is particularly striking is the lack of repetition in the developing universe. The fireball that begins the universe gives way to the galactic emergence and the first generation of stars. The later generations of stars bring to being the living planets with their own sequence of epochs, each differentiating itself from the rest. Biological and human history with their ever fresh expressions of creativity continue the differentiation of time from its beginning. Indeed all fifteen billion years form an epic that must be viewed as a whole to understand its full meaning. This meaning is the extravagance of the creative outpouring, where each being is given its unique existence. At the heart of the universe is an outrageous bias for the novel, for the unfurling of surprise in prodigious dimensions throughout the vast range of existence. The creativity of each place and time differs from that of every other place and time. The universe comes to us, each being and each moment announcing its thrilling news: I am fresh. To understand the universe you must understand me.

COSMOGENESIS IS STRUCTURED by *autopoiesis*. From autocatalytic chemical processes to cells, from living bodies to galaxies, we find a universe filled with structures exhibiting self-organizing dynamics. The self that is referred to by autopoiesis is not visible to the eye. Only its effects can be discerned. The self or identity of a tree or an elephant or a human is a reality immediately recognized by intelligence, even if invisible to senses. The unifying principle of an organism as a mode of being of the organism is integral with but distinct from the entire range of physical components of the organism. It is the source of its spontaneity, its self-manifesting power.

Living beings and such ecosystems as the tropical forests or the coral reefs are the chief exemplars of self-organizing dynamics, but with the term autopoiesis we wish to point not just to living beings, but to self-organizing powers in general. Autopoiesis refers to the power each thing has to participate directly in the cosmos-creating endeavor. For instance, we have spoken of the autopoiesis of a star. The star organizes hydrogen and helium and produces elements and light. This ordering is the central activity of the star itself. That is, the star has a functioning self, a dynamic of organization centered within itself. That which organizes this vast entity of elements and action is precisely what we mean by the star's power of self-articulation.

With such an understanding of the term, we can see that an atom is a self-organizing system as well. Each atom is a storm of ordered activity. This invisible power assembling the energy into a particular constellation is the atom's identity. A galaxy, too, is an autopoietic system, organizing its stars into a nonequilibrium process and drawing forth new stars from its interstellar materials.

Autopoiesis points to the interior dimension of things. Even the simplest atom cannot be understood by considering only its physical structure or the outer world of external relationships with other things. Things emerge with an inner capacity for self-manifestation. Even an atom possesses a quantum of radical spontaneity. In later developments in the universe this minimal dimension of spontaneity grows until it becomes a dominant fact of behavior, as in the life of a gray whale.

The question arises of the relationship between the autopoiesis of earlier and later manifestations. We need to ask about the sentience of the elemental realm, for we now realize that where Earth was once molten rock, it now fills its air with the songs of birds. And if humans bask in an astounding feeling for the universe, and the human arose from the elements, what can be said of the inner world of the elements? In our probing for understanding, we need to preserve both the continuity holding together an integral universe and the discontinuity enabling the universe to develop through a sequence of transformations.

The molten rocks in their original mode of being contained powers not present in contemporary rocks. Chemical processes common in the early Earth became rare as potentialities were activated and new epochs evoked. The movement from primordial molten rocks to mammalian consciousness is a radical one, certainly. But we must avoid regarding such consciousness as an addendum or as an intrusion into reality. The integrity of the universe must be respected; any theory that disregards the essential integrity of the universe dissolves the very basis for scientific knowledge. The universe is a single if multiform energy event; everything comes forth out of the intrinsic creativity of the universe.

Our interpretation speaks of the radical emergence of the qualities of the universe. The sentience of today's world is an ontological creation of the evolving universe. In former times it existed as a latent possibility; now it exists in its activated or historical realization. Because creatures in the universe do not come from some place outside it, we can only think of the universe as a place where qualities that will one day bloom are for the present hidden as dimensions of emptiness.

Even so, this interpretation of epigenesis—that qualities arise from emptiness, from the latent hidden nothingness of being—needs also to speak of the direct and intimate relationship between the powers of early Earth and the potential sentience in such powers. The rocks and water and air, just by being what they are, find themselves flowering forth within sentient beings. At the very least we can say that the future experience in a latent form is wrapped into the activity of the rocks, for within the tur-

76

bulence of molten magma, self-organizing powers are evoked that bring forth a new shape—animals capable of being racked with terror or stunned by awe of the very universe out of which they emerged.

COSMOGENESIS IS ORGANIZED by *communion*. To be is to be related, for relationship is the essence of existence. In the very first instant when the primitive particles rushed forth, every one of them was connected to every other one in the entire universe. At no time in the future existence of the universe would they ever arrive at a point of disconnection. Alienation for a particle is a theoretical impossibility. For galaxies too, relationships are the fact of existence. Each galaxy is directly connected to the hundred billion galaxies of the universe, and there will never come a time when a galaxy's destiny does not involve each of the galaxies in the universe.

Nothing is itself without everything else. Our Sun emerged into being out of the creativity of so many millions of former beings. The elements of the floating presolar cloud had been created by former stars and by the primeval fireball. The activating shock wave would have been ineffectual but for the web of relationships within the galactic community. The patterns of nuclear resonances enabling stable nuclear burning was not the Sun's invention—and yet all that followed depended upon this pattern of interconnectivity in which the Sun arose.

The universe evolves into beings that are different from each other, and that organize themselves. But in addition to this, the universe advances into community—into a differentiated web of relationships among sentient centers of creativity. The next chapters celebrate the reality of interrelatedness, especially within Earth and life. For here, to complete our introduction to these general themes of the universe's story, we will simply mention that relationships are discovered, even more than they are forged.

An unborn grizzly bear sleeps in her mother's womb. Even there in the dark with her eyes closed, this bear is already related to the outside world. She will not have to develop a taste for blackberries or for Chinook salmon. When her tongue first mashes the juice of the blackberry its delight will be immediate. No prolonged period of learning will be needed for the difficult task of snaring a spawning salmon. In the very shape of her claws is the musculature, anatomy, and leap of the Chinook. The face of the bear, the size of her arm, the structure of her eyes, the thickness of her fur—these are dimensions of her temperate forest community. The bear herself is meaningless outside this enveloping web of relations.

This sense of relatedness coming even before a first interaction, of a community reality at the base of being, characterizes even the earliest eras of the universe, even before the formation of natural selection pressures in the natural world. At this level quantum nonseparability governs activity. No two particles can be considered completely disconnected, ever. The particles in the fireball rise into existence in a direct and unmediated relationship with the rest of the particles in existence. This primal contact or togetherness of things represents the third systemic feature of evolution, through all the branches of its adventure.

We can see the value the natural world places on relatedness in the intricate mating rituals that have been invented. So much of the plumage and coloration and dance and song of the world come from our desire to enter relationships of true intimacy. The energy we and other animals bestow upon this work of relatedness—the attention we give to our physical appearance alone—reveals something of the ultimate meaning of communion experiences.

The loss of relationship, with its consequent alienation, is a kind of supreme evil in the universe. In the religious world this loss was traditionally understood as an ultimate mystery. To be locked up in a private world, to be cut off from intimacy with other beings, to be incapable of entering the joy of mutual presence—such conditions were taken as the essence of damnation.

A CLOUD OF ELEMENTS hovered, floated, slowly turned in its weightlessness far from the center of the Milky Way galaxy. In an indifferent universe, such a cloud would have been left to drift unchanged throughout eons of time, no significant transformations ever to take place, no intimacies evoked, all potentialities abandoned in their latent state.

In our universe, the originating powers permeating every drop of existence drew forth ten thousand stars from this quiescent cloud. To varying degrees, these stellar beings manifested the universe's urge toward differentiation, autopoiesis, and communion. And at least one of these, the Sun, managed to enter the deeper reaches of the universe creativity, a realm where the complexity, self-manifestation, and reciprocity at the very heart of the universe revealed themselves in a way transcending anything that had occurred for ten billion years—as an extravagant, magical, and living Earth burst into a new epoch of the universe story.

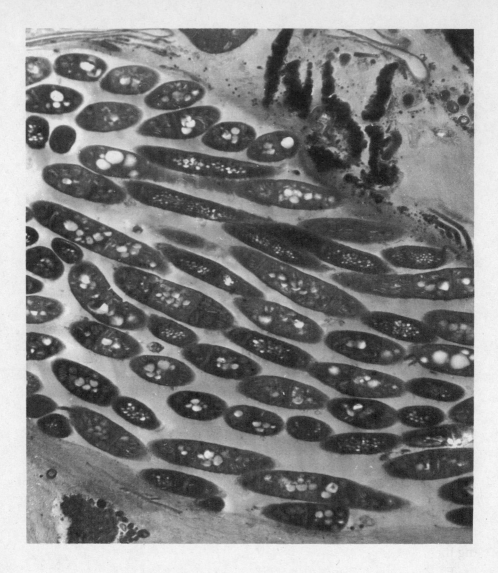

Photosynthetic Bacteria

5.
Living Earth

IN THE BEGINNING the universe flickered into a hundred billion clouds, out of which the galaxies constellated. In the spiral galaxies the flame of the beginning transmogrified into an epoch of supernova blasts, whose explosions stirred each spiral galaxy into a new concoction of elemental potentiality. In the elliptical galaxies, in the disturbed and irregular galaxies, in the globular clusters, and in intergalactic space, universe activity continued at its previous level, the pre-supernova capacity. Stars that had already been created maintained themselves. Nebular clouds of hydrogen and helium shone with ionized filaments. But none of these branches had reached the next depth in creativity at which spiral galaxies existed. The activities taking place in other galaxies also took place in spirals. But there were creative actions in spirals not found in any other place in the universe. The vanguard of creativity in this cosmogenesis was in the billions of spiral galaxies scattered throughout the universe. In such galaxies, things were brewing, novel forms were sprouting forth, new possibilities were presenting themselves.

Four and a half billion years ago in one spiral—the Milky Way—near the star coming out of a recent supernova explosion—our Sun—a disc of gases revolved. In and through this disc the universe entered into a new phase in the unfolding of its great adventure. The very existence of this disc had required five billion years of stellar labor. The elements composing this wispy shadow—the molybdenum, the titanium, the argon, the thorium, the iron, the neon, the fluorine, the calcium—all sparkled, each with unique properties, each a quantum personality with promises for activity beyond the imagination of the earlier universe.

As the carbon, phosphorus, beryllium, nitrogen, and all the elements created by the supernova collided, they stuck and formed small grains whose instabilities shaped the disc into at least ten bands. As centers of concentrations developed and swept even more of the matter into themselves, the spherical planets emerged that would later be called Mercury, Venus, Earth, Mars, Jupiter, Saturn, Uranus, Neptune, and Pluto.

In the beginning, each of the planets was molten or gaseous, due to the heat generated by the collapse of the presolar cloud as well as by the radioactivity of the elements. The radioactivity of the planets was a natural consequence of their violent birth. Elements forged in supernova blasts included the unstable isotopes, especially of such elements as carbon, uranium, thorium, and potassium. The heat from the breakdown of these atoms, combined with the frictional heat from the cloud's collapse, kept Earth in a burbling sea of molten rocks.

In a liquid, semi-liquid, or gaseous state, each planet rearranged itself according to the ordering established by the gravitational interaction. The heavier elements such as iron and nickel settled in the core, the middle mass elements formed around these, the lighter elements such as oxygen and silicon formed a layer still further out, and the lightest elements of all, such as hydrogen and helium, formed the outer layers. Each moment brought a new development in the composition of these planets.

The cloud left by the supernova had been drawn out of its equilibrium state by the gravitational dynamic and brought into the negentropic condition of a layered planet. When Earth reached the state where the heaviest elements were at the center and the lightest at the surface, the geological ordering would have ended had it been guided by the gravitational attraction alone. But breaking up this static order was the radioactive energy. The unstable elements created by the stellar explosion broke apart and released their energy. This heat, arising throughout the planets, kept them in a boil. Great currents of energy brought the materials from deep inside the planet to the surface, where they mixed and combined with the lighter elements there. Out of this planetary spume with its incessant chemical creativity, the atmospheres of the various planets were transformed.

The nine all had their hundred chemical gifts. They all brought forth the first mineral combination of the elements. These nine planetary centers of creativity revolved around the new central Sun. The great spheres churned with the searing heat from their collapse and from the time-released bursts of supernova energy. All nine began with the same elements, spun by the same star, burning with the same energy. Mercury, Venus, Earth, Mars, Jupiter, Saturn, Uranus, Neptune, Pluto—nine beings out of a supernova blast now left to their own powers to make their way into the adventure. But if this all began the same way, the destinies were vastly different.

O N MERCURY, VENUS, MARS, AND PLUTO, all geological activity slowly came to a halt. Within a billion years their geological boil froze. On Jupiter, Saturn, Uranus, and Neptune, the activity persisted without qualitative change. The storms taking place today are hardly different from what they were at the beginning of the solar system. But on Earth, such geological activity gave rise to continents, and to life, and continues to churn throughout the planet even five billion years after the beginning. What is the nature of Earth that this could happen and could continue for so long?

On Earth the universe could act with great subtlety whereas on Mercury, Venus, Mars, Jupiter, Saturn, Uranus, Neptune, and Pluto, the same degree of elegant activity was not possible. The dimensions of the other planets precluded such activity. The smallest planets—Mercury, Venus, Mars, and Pluto—specialize in formations of permanent rocks. Their sizes cannot generate enough gravitational pressure to break down the rocks that have already formed. As the ages went by, Mars, for instance, continued building ever more magnificent mountains, until finally the thick rock crust of the planet choked off the dynamism of the interior heat of radio-activity. A planet whose crust is stronger than its power to generate currents of materials is in its finished form.

Jupiter, Saturn, Uranus, and Neptune certainly had enough gravitational pressure to break down any rock formations. In fact they had so much gravitational strength no crust or rocks ever formed on any of them. Jupiter, Saturn, Uranus, and Neptune have remained churning balls of elemental materials in gaseous form. None of them advanced to the stage of creating stable continental formations. Though all the planets were permeated with the same universe activity, none was able to enter developmental pathways qualitatively similar to what took place on Earth because as whole planets they did not exhibit the elegant balance necessary for life and mind to emerge.

What did Earth have that the other planets did not have? Nothing spectacular—only the proper size enabling gravitational and electromagnetic balance. Nothing extraordinary—just a position with respect to the Sun that enabled Earth to establish a temperature range where complex molecules could be formed. These arrangements of matter enabled the solar system to advance to Earth's creativity. Had this particular amount of stellar material not congregated in such a size and at such a place, the solar system would probably have remained a lifeless place throughout its billions of years of existence. But such balance and such possibilities did emerge, and Earth became the advanced edge of cosmogenesis in the solar system.

Though Earth's birth has dimensions of chance associated with it, we begin to realize that a biophysical planet is no accident for the universe. It seems to be accidental that such a planet as Earth would today be revolving around our particular star; but it seems a probabilistic certainty that somewhere in the universe there would be a biophysical planet revolving around some such star. The universe, when given a chance, will organize itself into complex and persistent patterns of activity. Earth was the chance the solar system got. Earth was the gateway into a new and thrilling epoch of the universe story.

Vast nets of information had to be evolved. Carefully sedimented layers of biological accomplishments had to be maintained. A creative work that had appeared in none of the billions of elliptical galaxies was about to be undertaken. But in the beginning there were only these potent grains of dust gravitating around a fitfully burning young star. In the beginning there were only helpless planetoids colliding with one another, sticking together, smashing apart, cohering, growing, exploding under impact, reassembling, gathering mass in the storms of collisions to become, finally, a planet.

For several hundred million years Earth suffered the shock of collisions with meteors and planetoids. Just as an anvil when it is struck repeatedly by heavy blows from a sledge hammer will grow hot to the touch, so too Earth melted under the heat and boiled day and night for half a billion years. When the long period of collisions ceased, and Earth cooled just enough for ultrabasic rocks to solidify on the surface, plumes of glowing mantle rose up and scattered this solid ground into great chunks that once more melted down and sank into a boiling Earth. Chemical compositions forged in the interior of Earth were released at the surface so that each day the chemistry of Earth's air was new.

The early atmosphere of methane, hydrogen, ammonia, and carbon dioxide churned with an ionized turbulence more violent than anything it would again experience. Rainfall of a variety of liquids would begin high in the atmosphere and evaporate before reaching the molten rocks at the surface. Gigantic electrical storms, thunder that crisscrossed every point of Earth's surface, and immense lightning bolts all charged Earth for a hundred million years, propelling the planet through chemical gateways to a radically new branch of cosmogenesis.

Earth's groping into new worlds of possibilities was most intense where the surface cooled enough for liquids to remain and thus allow a triple interaction: planet-sized currents of molten matter boiled up from Earth's core to meet the turbulent action of the seas and the charged ions of the atmosphere. Out of this multidimensional cauldron, new beings emerged.

L IFE WAS EVOKED by Earth's dynamics, ignited by lightning. Not from a single branch of lightning, but a planetwide lightning storm stinging the oceans for millions of years. The word *lightning* fails us here. Even those of us who have been caught in a storm where a flash of light has split open an oak have experienced nothing in comparison to the lightning storms at the beginning. Today's lightning is only a faded memory of the earlier lightning. Perhaps we should speak of *ur*-lightning, the first lightning of the planet, a lightning that had never been here before and swelled with the supreme confidence of youth. Such unrestrained exuberance seethed throughout the clouds of our beginning. Out of such *ur*-lightning life came forth.

Each region of the universe then and now is permeated with self-organizing dynamics in latent form. These ordering patterns hide until the material structures and free energy of the region reach that particular complexity and intensity capable of drawing such patterns forth. A critical moment was achieved by Earth some four billion years ago in the cauldron of its early activity. Drawing upon the work of the star Tiamat who fashioned the hundreds of elements of Earth's materials, relying upon Tiamat's energy and the heat from the collisions in the new solar system, riding through the contoured pathways of gravitational and geochemical interactions, Earth took complexity to the extreme limit of inanimate forms, and in a searing lightning flash witnessed the emergence of a profoundly novel event—Aries, the first living cell.

Aries emerged from the cybernetic storms of the primeval oceans and found itself alone, in a sea devoid of any other life forms. But Earth would never again be the same. Had it not been lightning storms, but some other initiatory act, perhaps different self-organizing powers would have been drawn forth, with qualitatively different kinds of life. But the world-creating fact in our region of space-time—this region with its own particular history of galactic and solar and terrestrial development—the world-shaping fact is the emergence of Aries. It was Aries, in all its particularity, who came forth. It was Aries who would be for all time Earth's most ancient ancestor. In ways that we, four billion years later, are only just beginning to appreciate, it was Aries—in the shape of its body, in the processes within its life, in the ordering sequences of its information—who determined in a basal sense so much of the story that we have entered many millions and billions of years later.

Sun and Earth awakened Earth's life. The energy of the solar system transformed atmosphere and ocean, swirled into young seething Earth and erupted as lightning that initiated the proto-cellular chemical reactions. Life here was born in a lightning flash. Earth's life is lightning embodied and made flesh. Aries and all of its descendents were founded in a flash of lightning.

These new autopoietic structures, the living cells, were dependent upon the conditions of the universe for their origin. They could not have arisen without a particular initiatory sequence ultimately carried out by other beings. But once such new dynamic centers emerged, their very powers of self-organization enabled them to step to the task of maintaining their existence. They could resist encounters that pushed them toward extinction. In a similar way, the Sun had required the galactic density arm for its birth. But once the Sun's powers were evoked, it could endure disturbances without losing its basic identity. It could suffer the impact of millions of tons of interstellar matter, or the disturbances from the density arm that visited it every hundred million years. The Sun had the power to organize its presence against the sporadic onslaughts from the universe.

ARIES AND ITS immediate descendents were the most fragile autopoietic structures yet to appear in the story, and yet they were essential for the next advance in the story. They would dissolve under a pressure or a heat or an impact that would not even be noticed by previous autopoietic structures such as the star or atom or galaxy. But though more delicate than all those, cells brought autopoiesis, differentiation, and communion to new levels which made up for their fragility and enabled them to endure.

Aries' most impressive power of autopoiesis is that of memory. Cellular memory powers all of life, for nothing is more important to a living being than memory of the past. The first cells, the prokaryotes, still remember the early Earth by the very composition of their bodies. The seas are now different. The atmosphere is now different. The crustal composition is now different. But Aries' prokaryotic descendents living through Earth today still make themselves out of the same hydrogenous and carbonaceous molecules that were once prevalent in the early Earth. For four billion years the prokaryotic organisms have been remembering the composition at the beginning. Most impressive of all their feats of memory is to remember how they were created. They display this memory every time they create another version of themselves. These cells have the power to revivify that sequence of events that brought them into life. It is this power of memory that distinguishes the autopoiesis we call life.

In the turbulent creativity of the early Earth a spectrum of self-organizing structures was evoked, the vast majority of which lapsed back into chaos soon afterward. But one, or a family, or several radically different branches of families were called forth that had the power to reproduce the very sequence of activities that had called them forth. They "remembered" this elegant and complex sequence of actions, in the sense that they possessed the ability to bring this sequence into act. They were thus not dependent upon the universe to recreate the wild world that had brought them forth in the beginning. By themselves they could repeat the amazing

87

chemical act of creating life and could do so in a way that required only the slightest sliver of the energy that had been used in the beginning.

Dynamic centers possessing such memory promised to become deep and enduring features of the universe. Though fragile, though liable to destruction and change in an infinity of ways, they could nevertheless perform an aboriginal magic that world enable them to pervade the world: they could swallow a drop of seawater and spit out a living version of themselves. What did it matter if individual cells were vulnerable to the strike of lightning or the sudden upwelling of lava? In a matter of moments cells could recreate the miracle of fashioning living beings. If enough of them were created all danger could be sidestepped by having living cells everywhere.

Besides these new powers of autopoiesis, cells exhibited a new depth of differentiation as well. Throughout the whole of the early seas, cells gave birth to new versions of themselves, but they did so in a way that produced offspring slightly different than themselves. Once every million births, a cell was created that was new. In the beginning, protons were formed by the trillion, in identical copies of each other. So too with the emergence of the later autopoietic structures of hydrogen and helium. But with cells, such identity did not hold. Offspring were different, slightly changed in the form and functioning of some of their five thousand protein molecules, slightly different in their ability to interact creatively in their exchanges of mass and energy.

This phenomenon—*genetic mutation*—is a primal act of life. Mutation explains a great deal of life's story in much the same way that gravitation explains a great deal of the galactic story. Both explain but neither can be explained. Each is a primal act. Each refers to ultimacy. To say a mutation is a primal or ultimate act is to say that it is an irreducible activity at the heart of the unfolding universe. A mutation cannot be explained in itself any more than the strong nuclear interaction can be explained in itself. As with the consideration of any primal act, the very most we can hope for in our investigations is a deepening understanding of its significance within the story of the universe.

Aries' immediate descendents, the early prokaryotes, proliferated and differentiated so that in time they filled the seas. They fed on the exotic, energy-rich chemistry created by the turbulence of the hot early Earth.

Each prokaryote had from the beginning relied upon a diet of diverse chemical creations for its continued existence. But as the Earth's turbulence slowed down, the production of such compounds slowed down as well, and with prokaryotes expanding numerically at an exponential rate, and thus consuming the feast ever more quickly, the first cells headed toward a wall. At the very least their numbers would have to be drastically cut back until their food needs just matched the reduced ability of the now calmer Earth to generate such chemically rich compounds. It was possible that this great die-off might not take place soon enough; that some essential and scarce chemicals would be driven to virtual nonexistence, leaving the few remaining prokaryotes on a dying planet that had no food and that had lost its previous power of evoking living cells out of the elements.

Mutations sidestepped disaster. New forms of prokaryotes appeared that could feast on the body parts of perished prokaryotes. Others appeared that could consume the compounds produced by prokaryotes. These new micro-organisms took what was waste for one kind of being and discovered that it could become food for another kind of being. These forms with their slightly different capabilities wove the first communities of life. Circles of involvement arose. A decayed body became food for living cells, and thus was transformed into either a living cell or waste that was food for another, which in turn became part of that cell or passed out as waste. In this way, webs of relationships were created that were new in the universe story.

Another mutation appeared that is one of the greatest acts of creativity in the four billion years of the living Earth. This mutation took the porphyrin ring—an earlier discovery that proved to have the delicate ability to handle individual electrons—twisted it in conjunction with other chemicals and discovered a molecular net with the power to capture photons in flight. This mutation had the ability to convert the energy of a particle rifling through the air at the speed of light into the molecular structures of food. Suddenly, in at least one cell, in Promethio, a new intimacy was established between Earth's living surface and the radiant energy from its central star.

THE EMERGENCE OF Promethio with its photosynthetic ability to interact creatively with the photons from the Sun illustrates as well as anything else the wild wisdom at the heart of the universe story. The creativity of

Promethio is an act of unsurpassable elegance—and from a being one millionth of a meter across, weaving a molecular power astounding to behold, and doing so without a brain, without eyes, without hands, without blueprints, without foresight, without reflective consciousness. Mutations appear, seemingly random events in an ocean filled with random events. A community of beings is left to die on a planet without food, wandering around in search of a way out. A mutation surfaces bestowing the power to feed on the Sun. Life powers forth billions of novel cells, tossing them into the great adventure to see what might come of it. The cosmological power of differentiation explodes with a trillion new pathways. Let them all enter existence. Perhaps one will result in a thrilling new complex community where the enjoyment of every participating being will be deepened.

No individual prokaryotic cell strove to invent a molecular process that would capture the particles of light. Their inner-felt experience could never contain such a complex mental idea and would be minimal at most. We can imagine a cell experiencing in some dim manner the dull ache in the center of the autopoietic activity when nutrients were absent. There would be the simple striving to capture what was required for another burst of life. But all considerations of the Sun, of photons, or of new metabolic pathways would be light years beyond the conceptual capacities of the cell's subjectivity. All such considerations were opaque to cellular experience. But within this capacity Promethio and the other bacteria also had access to cosmos-creating powers of differentiation. That alone within a limited interiority was enough to invent a lifesaving strategy that was quickly communicated to disparate communities throughout the terrestrial ocean.

Cells not only have the power of memory, they have the ability to share their memories among themselves. Promethio, as the first cell mutant to discover photosynthesis among its many powers, made its new power available to some of those bacteria with whom it had intimate contact. Promethio shared this epochal power by creating a small tube connecting it to another prokaryote, inserting some of itself into the tube and sending this gift into the other cell. The material that was inserted—the "episome," the "plasmid," the "replicon"—this material gave the receptive cell the power to duplicate the feat of photosynthetic capture of light particles. Now it too had "learned" how to photosynthesize. It too "remembered" how to feast on the Sun.

The material transmitted contained in a physical way all the information the new cell needed to photosynthesize. This ability to store information in matter is what enables a cell to remember past experiences, to recreate them in the present, and to share such memories. This richly coded matter—the genes of the deoxyribonucleic acids—by carrying in its structure the vital information, enabled each cell to share with any other cell what it knew of the world. Any cell receiving crucial information suddenly has the power it needs to live and multiply. Mutations that are vital thus become widely present throughout any community of life in which they arise.

When a later bacterium absorbs the Sun and creates food, a new intensity of communion in cosmological epigenesis has been attained, for so much of the universe comes together in that one event. Both the bacterium and the Sun are composed of the elementary particles of the first Flaring Forth that took place some ten billion years previous to their time. Both are composed of the hydrogen atoms created at the end of the fireball. Both the Sun and the prokaryote are composed of the elements created by the supernova Tiamat that exploded half a billion years ago in their common past. Materials from all these remote events are present in the new activity. In that sense the past is there, alive in the present.

But with the emergence of cells and their ability to remember the past, and to share memories of the past, a new intensity in the presence of the past is attained. When this later bacterium photosynthesizes radiant energy, its activity mirrors that first photosynthetic act of Promethio occuring millions of years earlier. The new cell has the same information from that first mutation event, and thus can absorb the sunlight now.

The genetic memory of the cells enables single events to take on supreme significance, and thus autopoiesis reaches a new level of meaning. If out of a trillion mutations, just one brings a vital adaptation into Earth, just that one mutation has a chance of becoming a pervasive reality throughout the entire living community and for billions of years into the future. In such a case, the essence of this one mutation event would be repeated endlessly, called upon by trillions of future beings. Its power to evoke beautiful life, that surfaced radically just once, can become a common power throughout vast communities in the future.

The universe becomes present to itself in this new way because life strives to capture and save its significant breakthrough experiences. If one event in a quadrillion reaches a new level of creativity, that singular moment has the possibility of being remembered and shared. One by one such creativity is saved. By accident some rare and beautiful breakthrough events are forgotten, but the aim of the cells is to remember such effective creativity. The sum total of all such significant historical events cannot be contained by any particular bacterium. The DNA of each is not long enough to contain all memories that might some day be valuable. Instead, the memories are dealt freely among the living bacteria so that the irreplaceable learning is held by the community as a whole. It is thus the entire community that probes, creates, discovers, learns, develops, and remembers..

J UST AS LIFE'S memory binds together events that are separated in time, so too, life binds into a vital relationship different levels of being. The early Gaian system of churning magma, charged atmosphere, surging seas, and crystallizing minerals gave birth to the first living cells; these cells in turn gave birth to a new era of the Gaian system as can be seen most clearly by telling the story of carbon and oxygen.

The atmosphere of the early Earth became increasingly dominated by carbon dioxide as volcanos released carbon locked up beneath the surface into the air. At first, the air's carbon concentration was unaffected by the early forms of life, for they did not interact with it. But mutations arose with the power to draw carbon dioxide into their autopoietic cycles. Since such bacteria could flourish in comparison with forms unable to use the carbon in the air, this new ability would characterize increasing numbers of bacterial communities. Once connected to the cycles of the living, the carbon of the atmosphere would steadily decline as it was transformed into living bodies. Not only the atmosphere but the oceans, too, would be changed, for they would not be able to absorb as much carbon from the air. Even the rocks would be altered, for once the bacteria died their body parts sank to the ocean floors to be pressed into limestone which would be drawn into the convection cells of the magma sinking into Earth's center. Thus, through the evolution of life, the three macrogeological cycles of atmosphere, hydrosphere, and lithosphere were changed in their chemistry.

In the beginning, living cells one-millionth of a meter across had been evoked by the creative interactions of the three macro-systems of the early Earth. These cells existed in a parasitic or derivative sense, infants in a world of adult planetary transactions. But once the cells mutated to forms of life that could interrelate with these cycles, life as a whole entered planetary dynamics as an equal partner. Thus activated, the biospheric dimension of Earth's epigenesis immediately altered the terrestrial unfolding. Earth's adventure became a conversation among the hydrosphere, lithosphere, biosphere, and atmosphere.

Though negligible in the beginning, the effects of the biosphere on Earth's story soon became massive. An atmosphere with carbon dioxide is able to retain the thermal radiation from the Earth's surface, so as cells pulled carbon out of the air, the temperature of the entire Earth declined. By 2.3 billion years ago, Earth sank into a glaciation that spread immense ice covers across the continents. The continents themselves had been stabilized 3 billion years ago, with the major continents established by 2.5 billion years ago. But now, made heavy with ice packs, they entered another phase of their development. The cold cracked the rocks, enabling seawater to penetrate the continental shelves and dissolve their minerals. Thus, the first vast ice age, with all of its effects on the seas and the continents, is a result of the planet having activated its biospheric dimension.

YET ANOTHER BIOSPHERIC invention destroyed the Archean, the first eon of Earth's life.

The earliest cells satisfied their hydrogen needs most easily by using the supply of rich nutrients created by the turbulence of the primal Earth. As these nutrients were used up, life mutated into blue-green bacteria that took hydrogen from the seas. But to cleave hydrogen from the water molecule is to send oxygen into the cybernetic networks of the planet.

The powers of oxygen are among the most formidable of all the elements. Oxygen is in perpetual need of electrons and will rip them out of previously stable compounds, leaving instead an electron-hungry radical that will repeat oxygen's theft elsewhere. As oxygen was released into the geological cycles it transformed the lithosphere and atmosphere first. It invaded the rocks and either removed elements from them or cemented itself to them, thereby changing the chemical nature of the land. Once the surface had been transfigured, oxygen grew in strength in the atmosphere and disrupted its ammonia, its carbon monoxide, its ancient and previously irreplaceable hydrogen sulfide. As its work in altering the nature of Earth's atmosphere neared completion, it hovered over the land and leapt upon any new upwelling of molten rocks from the interior, quickly changing their chemistry as well.

By two billion years ago, oxygen's radical transformation of the lithosphere, hydrosphere, and atmosphere had been carried out. The biosphere could no longer avoid facing this profound disruption. At first, cells had been able to ignore oxygen, since its concentration was so low. Later, cells had been able to hide in the water to keep away from the presence of oxygen. But after a billion years every arena had been invaded by oxygen and the very source of this oxygen—the biosphere—had to deal with its presence.

The destruction that followed was one-sided. Oxygen began by eliminating the food supply. Carbohydrates rich in energy, desired by all bacteria, were rendered chemically useless as oxygen broke down the richest molecules. Oxygen penetrated into the food the instant it was created and made it inedible, changing it into particles of dust. Oxygen also dealt directly with the cells. The fragile membranes were attacked by the oxygen until the cellular cytoplasm broke free, leaving the cell to expire into dispersing droplets of its own components.

Oxygen also slid through the cellular membranes and took apart the enzymes, leaving the cell helpless to perform its life-sustaining tasks. Oxygen entered and changed lipid molecules into slowly burning flames. In groups of ten trillion, oxygen attacked the cell's DNA nerve center, those complex and information-rich nucleic acids, consuming electrons and tearing apart structures. Where there had been a vibrant living cell, there remained only a disconnected and chemically meaningless scrap heap, all that subtle and elegant information evaporating as if it had been inscribed on a papyrus roll now in flames.

Where once there had been a vast multilevel community of prokaryotic beings in intimate relationships, there were left scattered remnants still under attack from a lifeless enemy, their own creation. With each passing day the richness of the Archean world faded still further. The first great eon of the biosphere's history ended in catastrophe.

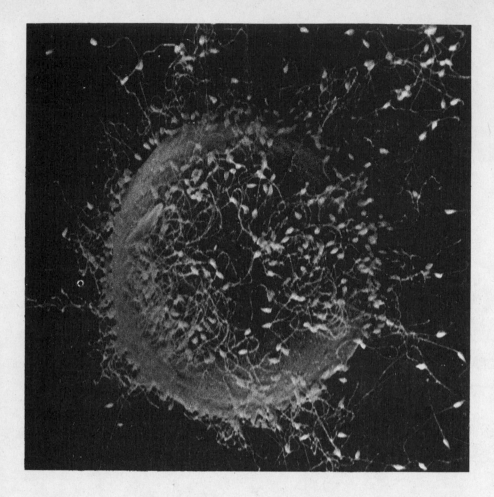

Sperm and Egg, Sea Urchin

6.
Eukaryotes

OUT OF THE supreme radiance, the trillion protogalaxies appeared and gathered together into full-sized galaxies, clusters of galaxies, clusters of clusters, vast sheets crisscrossed with filaments of giant galaxies. Out of the billions of spiral galaxies rich with the supernova's generosity at least one planet brought forth Aries from its boiling rocks, the first living cell, a being both fragile and powerful. The first continents began to stabilize, and by 2.5 billion years ago half of them had risen out of the oceans. The atmosphere was brownish orange, consisting mostly of nitrogen mixed with carbon dioxide and methane gases. The oceans, also brown, were the locus of life, though some of the prokaryotes survived along the edges of the continents, clinging to and wearing out the rocks whose nutrients were released into the seas. There were no ears, but had there been, they would have heard the sounds of waves, the winds, the volcanic emissions, and the bubbles of the gases escaping from life. There were no eyes on the planet, but had there been, and had they watched for a billion years, as continents floated through the oceans crumpling into each other and joining together for a while, as volcanoes spouted, as the air and water slowly changed from brown to pale blue, they would have missed the real drama taking place. For Aries and all her prokaryotic descendents, too small for any eye to see, whose activities had changed the nature of the hydrosphere, atmosphere, and lithosphere, had been set on fire from the inside and were perishing throughout the planet.

Perhaps similar crises have happened elsewhere in the universe. Perhaps such crises have taken place in billions of other worlds, with a great variety of resolutions. Certainly we can imagine this experiment called a biosphere ending: microscopic beings, without eyes or brains, without hands, now having to deal somehow with such a dire situation. Just by being what they were, they had set themselves on fire in an oxygenated atmosphere. One can imagine catastrophic failure. Perhaps failure was the result on a billion planets. If so, such centers of activity would have simply disintegrated. The pale air and sea would have remained brown for a time, and then become transparent. The volcanoes would have continued for another few billion years. The windswept continents would have continued to collide and break up, collide and break up.

INSTEAD OF FAILING, life mutated. A cyano-bacterium appeared, Prospero, that could deal with this oxygen, this powerful element that was tormenting all life. As was discussed in the story of the supernovas, it was this very obstacle that made it possible for a creative advance to take place. For the bacterium Prospero not only survived, it invented respiration, the power to deal with oxygen. This alone gave it more than ten times the energy of any other cell.

Suddenly this cell Prospero appeared, and magically transformed a curse into a blessing. That which had been killing life now enabled life to burst with energy. Capable of exploiting that very element which every other cell was avoiding, Prospero soared into the most pervasive form of life. This cyano-bacterium had a seemingly infinite supply of chemical energy. It exploded into every region of the oceans. No other form of life could compete with it for nutrients, for Prospero was the complete cell. It got its energy from the Sun and its hydrogen from the water and its carbon from the atmosphere. And it powered its activities with the controlled combustion made possible with oxygen. Prospero had all the memories required to make everything it needed in the way of proteins and nucleic acids, and it had easy access to all the chemical materials necessary for its self-assembly. No wonder that Prospero proliferated madly and splintered into many different species.

Life had crossed a threshold. As cyano-bacteria continued to draw their hydrogen from the sea, the oxygen continued to build up, making it ever more difficult for other forms of life to persist. As ozone, the outer layer of atmospheric oxygen, thickened, the most energetic photons from the Sun were refused admittance to the sea below. Earth thus ended the billion-year epoch characterized by the abiotic creation of organic compounds in the seas. Those very conditions that had enabled the seas to become nutrient-rich were now gone. And hyper-energized by oxygen, Prospero's descendents increased the oxygen density hundreds of times beyond what the earlier Earth could have endured.

When oxygen concentrations rose above 21 percent, even Prospero spontaneously erupted into flames. But near 21 percent the combustion of respiration was still manageable, and Earth's complex cybernetic system of biosphere, lithosphere, hydrosphere, and atmosphere stabilized itself just below the level of spontaneous combustion.

The cosmogenesis of Earth and universe proceeds by establishing stable regimes that are then broken apart by the activities of the parts, thus leading to new stable regimes with new members included in the new dynamics. In the best of circumstances, the succeeding regimes are characterized by an enhanced diversity, a more extensive autopoiesis, and a richer interrelatedness. But at times, given the inherent turbulence of the universe, these later regimes are ripped asunder with no further development. The unfolding of the universe proceeds with reversals and surprises, sporadically advancing in various directions and then as suddenly collapsing.

The transformation of Earth under the trauma of oxygen is homologous to the transformation of the star Tiamat during the trauma of hydrogen exhaustion. In both cases, a situation far from equilibrium evoked a new being—Prospero-like bacteria in the case of Earth; carbon nuclei in the case of Tiamat. The appearance of these new beings completed the destruction of the former regime and established a new dynamic dissipative structure. Earth with an atmosphere of 21 percent oxygen is a complex interweaving of activities that is qualitatively different from an earlier Earth at 0.5 percent oxygen. So too, the star that is intense enough to cook the nuclei of carbon atoms is a different dynamic dissipative structure from the former stellar system still stable in its production of helium nuclei.

Earth through its complex interactions of atmosphere, lithosphere, biosphere, and hydrosphere established a new system. The biosphere initiated the breakdown, but it was Earth as a whole complex system that established a new harmony. The biosphere is but one of the subsystems involved in the oxygen, carbon, and other mineral cycles. For instance, carbon is removed from the atmosphere by the hydrosphere as well as by the biosphere. Rain absorbs carbon dioxide and changes it to carbonic acid, which reacts chemically with the rocks to release bicarbonate ions into the seas. The consequent carbonate sediment layers the ocean floors as they slide under the continents and sink back into Earth's interior. The cycle is completed when the absorbed carbonate materials are released again through the volcanoes as carbon dioxide. Such geochemical cycles existed before life emerged and persisted afterward, though altered by the presence of the oxygen from the new biospheric cycles. Only when the entire complex of these interconnected cycles established a new system of stability among themselves was the trauma of oxygen resolved.

The creation of new stars is not possible except in spiral galaxies. The creation and dispersal of the elements is not possible except in supernova stars. The universe needs special domains such as spiral galaxies and supernova stars for these creative advances to take place. So too, the creation of complex, advanced living beings is not possible in globular clusters nor on asteroids, nor in the systems of the earliest Earth. What is required for the emergence of advanced forms of life is a planet far from chemical equilibrium with an enormous sea of potential chemical energy. Aries' prokaryotic descendents initiated a process that broke the peace of the early Earth, disrupted the stable communities, and led to a zone intolerable for all early forms of life. Even so, this was a zone wherein Earth could flower in a way previously impossible. The creativity of the prokaryotes led to a wildly energetic geochemical dissipative structure whose new possibilities would not be fully plumbed even after another two billion years of creativity.

The eukaryotic cell, the first radically new creation within the oxygenated Gaian system, is the single greatest transformation in the entire history of Earth, only overshadowed in significance by the emergence of life itself. There are two basic eras of life: the prokaryotic era from four billion years ago to two billion years ago; and the eukaryotic era from two billion years ago onward. The eukaryotic structure opened up a reign of biological creativity bringing forth novelties unimaginable in the Age of Bacteria. Yet it was the bacteria that made it all possible.

Two billion years ago, Viking, a mutant form of the hyperenergized Prospero line, discovered a gruesome approach to life. Viking attached itself to a larger prokaryote, Engla; drilled holes in the cellular walls; invaded; and feasted on Engla's insides even while it employed Engla's DNA to manufacture its own proteins and nucleic acids. When Viking had gotten the food and information it needed, it multiplied until the host Engla exploded. Fired with a controlled combustion, a bacterium such as Viking would enjoy great success. Possibly, the ease with which Viking could use Engla's genetic material enabled it to slough off some of its own genes. In time Viking became dependent upon Engla's DNA. Thus it came about that without Engla and its necessary genetic information, Viking could not survive. Earth's first symbiotic relationship was under way.

Given the great variety of Englas and Vikings, plus the millions of additional mutations taking place as their relationship developed, the spectrum of resolutions for this tense interaction would spread between two extremes. At one pole, the most aggressive Vikings would finish off the host cell with great relish. Once all local Engla cells were consumed such Vikings would then perish, too, for lack of genetic materials. At the other extreme would be those Engla cells capable of destroying the invading Vikings. Such would be able to rid themselves of these lethal parasites but would then find themselves suffocating in the ever-increasing oxygen. Thus if either a Viking or an Engla defeated the intentions of the other, it too collapsed in failure. Viking and Engla, natural enemies, in the end embraced the symbiotic relationship that was fast becoming their only path into the future.

The first eukaryote might have emerged 2 billion years ago. We have fossils of eukaryotes from 1.4 billion years ago. Through the mystery of communion, sometime between 2 and 1.4 billion years ago, Viking and Engla created a vibrant new alliance of life. Viking fed on internal constituents of Engla, but restricted its feeding to Engla's waste products. Even better than simply surviving this infestation of Viking cells, Engla discovered two life-saving benefits. First, Engla could now survive in an oxygenated environment. The oxygen, instead of setting Engla's insides to a slow flame, would be quickly swallowed by the Vikings who gathered at the membrane edge within Engla. Second, with the intense energy production of these Viking cyano-bacteria, the host Engla could now enjoy powers it could never have generated by itself.

Newly vitalized, the Viking-Engla cooperative entered the adventure of life so successfully that its form became predominant throughout the oceans. For the next two billion years all major transfigurations of the life world would use this same form, this cooperative, this system of intimate interrelationship.

Starting as enemies, one a lethal presence to the other, Viking and Engla soon became so intimate they grew to depend upon one another for life. Circles of nutrients that had formerly included the world outside the cell were now contained within cellular walls. The waste of Viking was the food for Engla, and the waste of Engla became the food for Viking. The genetic material necessary for their common metabolic system was divided up between the two previously independent cells. If the Vikings within were to die for some reason, Engla would as quickly perish. Even the subtle and delicate process of creating new versions of themselves eventually proceeded as a single act. Their rhythms of life enmeshed with each other.

To call this organism a Viking-Engla cooperative underlines the symbiotic origin, but fails to appreciate the new individuality, the new identity. Even the genetic material of each was reduced somewhat as each symbiont came to rely upon the other. Viking in time evolved into the mitochondrion, an organelle that would be found in all complex cells for the next several billion years. Perhaps we can refer to the first great emergence in the oxygenated Earth as *Vikengla,* removing all reference to its previous life as two separate beings.

Earth had evolved to a far more energetic world than in the Archean, life's first eon. Eyes that could respond to infrared and microwave electromagnetic radiation would have seen the entire planet aflame with processes of oxidation. Volcanoes blasted into the atmosphere. Continents were ripped apart, and lava oozed across the surface. The rock's chemicals transformed as they mixed with the oxygen created by the complex burning Vikengla organisms. And the hyped-up energetics of Earth as a whole had now brought forth in Vikengla a microcosmic energetics of corresponding intensity. And this was just the beginning.

In a shallow pool along the ocean's shore at the edge of a continent that stretches thousands of miles, barren of all life, barren of all topsoil, just the angular blue-gray faces of granite all the way to the pale blue horizon, a diverse population of eukaryotic cells gathered. Two billion years earlier, the water had been rich in chemically energetic compounds evoked by the lightning storms and the cosmic radiation. But the storms had long since ceased and the oxygenated atmosphere now shielded the oceans from any intense radiation. So there was a population of life, a new form of cells, a nutrient-poor sea, a hyper-energized complex being corresponding to the flame of Earth's system.

LIFE MUTATED. A new cell, Kronos, brought forth a novel strategy. Kronos consumed its living neighbor. Unable to feed its flame by using the Sun or the sea's chemistry, a mutation in the Vikengla lineage broke with life's ancient traditions and swallowed a living creature while it still throbbed with life.

Roughly estimating this event as occurring one billion years ago would mean that life required three billion years to muster the audacity to live in this way. The ancient customs of chemoautotrophy (eating chemically rich molecules) and phototrophy (eating the particles of light from the Sun) or detritotrophy (eating the decayed bodies or wastes from other creatures) were pushed aside by this exuberant, unpredictable, high-strung eukaryotic mutant. Heterotrophy was born.

If life had proceeded with amazing versatility and power for three billion years without requiring that one creature devour another while it yet lived, heterotrophy is surely one of the freak mutations of life's journey. It need not have happened. But with the emergence of this new strategy of swallowing a cell whole, with the appearance of the primordial mouth—one certainly without teeth yet, although these would be quickly added if a fleshy, toothless hole was already enough to find one's way to undreamt-of food supplies—Earth's adventure turned onto yet another new branch with its own particular pathways into power and beauty.

Would Kronos have any awareness of what it was doing? Its autopoietic powers would have included minimal discernment. Any primitive eukaryotic cell would, for instance, be able to detect a temperature gradient, turning itself toward warmer regions. It would possess a limited ability to sense nutrient densities and orient itself to their thickest direction. Possibly Kronos would be capable of registering the presence of living cells. But in any event such a primordial creature certainly had no abilities of foresight, of making plans, of imagining the future. With its restive and stimulated dynamism, it jumped the tracks of history and brought Earth's adventure into the unknown, all without the least reflection on what it had done.

In its future, pale blue skies would shriek with the death terror of pteranodons seized by the quickly stabbing rows of knifelike teeth that lived in Kronos's descendents' mouth. Springboks would learn to eat with their ears ever attentive, lifting their heads silently at the slightest fluff of sound, the doe eyes perfectly still with fright, then exploding in a zigzag escape from the leap of a great cat who had learned its hunting skills from a long line of predators brought forth by the ancestral Kronos. Black eagles soaring with talons outstretched, orcas circling and confusing before shredding the great whale and darkening the seas with its blood, bats whipping through the night to devour thousands of churning insects—none of these would have trembled forth had not Kronos dared to probe this path.

When Earth took on heterotrophy it recoded its basic structures. Kronos absorbing whole a living cell is not just another instance of an autopoietic system obtaining energy for its continued existence. It is certainly that. But heterotrophy, the act of eating, is one of macrophase creative dimensions. A dark protostellar cloud had imploded to become a star with bands of matter swirling about it. The third band boiled with magma, showered itself with a million-year lightning storm, and brought forth pieces of itself moving under their own power. This third planet, churning in the great circles of hydrosphere, biosphere, lithosphere, and atmosphere, drove these gyres into new shapes and intensities when it took the daring step of eating itself alive.

The predator-prey relationship gives living populations a new complexity, turning them into ecosystems. A photosynthetic autotroph orients itself primarily with the sun and with a few minerals it requires from the seas. It has no vital need to attend to the activities of other living beings. Detritivores, since they feed off the waste and decay from other creatures, break the complete symmetry of the phototrophs, for they cannot thrive in simply any region of the ocean, but only in those where other organisms reside, and sometimes only in those regions populated by organisms of a certain kind.

The first heterotrophs pushed relationship much further. They began to specialize in certain forms of prey who in turn coevolved—who adjusted genetically to their role as the prey of a particular kind of predator. The form, functioning, and biochemistry of the prey began to reflect the form, functioning, and biochemistry of the predator, and vice versa. The predator heterotroph no longer lived in the symmetric world of the autotroph whose relationships with any one creature were the same as those with any other creature. Rather, the predator was vitally interested in one direction above all others, and in one form above all others, that of its prey.

Through its series of mutations, Kronos discovered the most effective metabolic pathways for digesting its prey. Such power, captured in its chromosomal DNA, became ever more pervasive throughout the Kronoid gene pool and thus Kronos became ever more attentive to the life of its prey. In a complementary manner, the prey coevolved by remembering those mutations that enabled its gene pool to survive. If no such spontaneous powers came forth the prey would collapse into extinction. In such a situation, Kronos would either discover a new form of prey or follow it into extinction as well; in any case, the relationship would be over.

Only those predator-prey relationships that coevolved into viable forms survived and propagated their patterns into the future. On the basis of such enduring symbiotic relationships the haphazard encounters of the Archean eon evolved into the ecological, systemic, patterned communities in the Proterozoic seas. It was out of the intimacy of such ecosystems that the third great surprise of the Proterozoic took place.

AFTER THE CREATION of Vikengla, the first eukaryotic cells, and the daring of Kronos with its heterotrophic nourishment, life mutated into Sappho, an organism with the advantage of meiotic sex—a shocking and seemingly unnecessary extravagance.

In their tumult of consuming living beings, some Kronos eukaryotes found themselves possessing whole living beings within themselves that refused to be digested.

Meiotic sex had its origins when Kronos was able to digest the cell walls of its prey, as well as some of the prey's cytoplasm, but not the prey's nucleus harboring its genetic treasures. Stymied in its attempts to feed,

Kronos would rest. But a vast transformation had taken place. Now Earth's adventure expressed itself in a single cell with two nuclei. Each of these nuclei was rich with genetic information representing billions of years of selectively stored memories. Each nucleus was capable of organizing matter and energy into a vital living creature. Earth had spawned something new: a being with two nerve centers.

A vast variety of experiments were undergone. In the majority of cases of consumed nuclei, the commands of the new nucleus would be dysfunctional. The cell would recognize and respond only to the cybernetic commands of its original nucleus. But there would also be situations when the alien nucleus possessed enough similarity in genetic language to evoke effective responses in the cell. In such a strange situation, with two functional centers of intelligence, the cell as a whole might be unable to function properly and would perish. In other experiments, the alien genes might be carried along without much effort, harmless stowaways. But in some cases, perhaps only in a very small number of situations, the new nucleus might have a profound relationship to the original nucleus, with a surprising harmony springing forth as a result.

If the consumed prey had been coevolving with the Kronos predator for a long enough period, the prey's nucleus would possess the power to produce those enzymes and amino acids especially desired by the predator Kronos. It was for just such elements that Kronos had pursued the prey in the first place. But if now the new nucleus could establish itself in a viable manner within Kronos, this new cell would have sudden access to a great body of fruitful information that would enhance its own adventure in life. It would possess twice as much genetic information as before, with a new set of chromosomes that complemented in various synergistic ways the original information.

How many trillions of beings perished in this enterprise? How many millions of mutations had to spring forth before a harmonious communion between the nuclei and their actions could be established and coded into the chromosomes? In the fullness of time, and after building on all previous experiences, the sexual eukaryotic cell Sappho finally came forth. Sappho possessed two nuclei. More precisely, since the nuclei had become intrinsically related to each other, we say that Sappho possessed a double or diploid nucleus, a nucleus that contained two complementary sets of

106

chromosomes. This diploid nucleus functioned as a symbiotic whole through all of the needs of the new cell except when it came time to divide and propagate.

When Sappho experienced the urge to reproduce itself, ancient memories of original cell division returned. First it scrambled the two strands of the nuclei so that parts of one changed places with parts of another. Each half of this shuffled diploid nucleus then separated and formed itself into a single cell. Each of these new cells had its own "ancient" nucleus, a nucleus having only a single set of chromosomes.

These special cells of Sappho, call them Iseult and Tristan, were then released by Sappho into the currents of the enveloping ocean. They were cast into the marine adventure, with its traumas of starvation and of predation. Able to nourish themselves but no longer capable of dividing into daughter cells, such primal living beings made their way through life until an almost certain death ended their three-billion-year lineage, all their memories perishing in their demise.

But a different end to the journey could also take place. A slight, an ever so slight, chance existed that a Tristan cell would come upon a corresponding Iseult cell. They would brush against each other, a contact similar to so many trillions of other encounters in their oceanic adventure. But with this one, something new would awaken. Something unsuspected and powerful and intelligent, as if they had drunk a magical elixir, would enter the flow of electricity through each organism. Suddenly the very chemistry of their cell membranes would begin to change. Interactions evoked by newly functioning segments of her DNA would restructure the molecular web of Iseult's skin, so that an act she had never experienced or planned for would begin to take place—Tristan entering her cell wholly.

This act that they had entered upon so rashly would lead to both of their deaths; or perhaps we should say to their rebirth, in a new form. Tristan's tumultuous entrance is followed by the dissolution of his cell membrane, and its absorption into Iseult's cytoplasm. Now Tristan's naked DNA is free to advance upon her own. Both genetic strands stretch themselves out in long undulations, with each laying itself snugly into the alien though deeply familiar contours of its partner. The new couple quickly creates a molecular membrane that curtains them away from the rest of the cell.

Sappho is reborn. But this is a new Sappho and in its novelty is the foundation of life's radically accelerated unfolding. The proto-sperm Tristan cell was the creation of one Sappho and the proto-ovum Iseult cell the creation of a different Sappho, each with its own lineage of potent memories. The new Sappho that they have created draws upon each of these lineages. The remembered experiences of one ancestral line might differ significantly from those of the other branch. Out of this new combination, synergistic novelties can arise in a single generation that would otherwise require many thousands of prokaryotic generations to develop.

Meiotic sexual reproduction of a new Sappho cell creates a relationship with a particular branching into the past, the ancestral tree. As opposed to bacterial sex, where the daughter cell is derived from a single parent cell, which in turn was derived from a single cell, meiotic sexual cells enable an entire branching network of the past to become present in the genetic activity of the new Sappho, advancing the interrelatedness of cosmogenesis.

The subjectivity of Sappho advances as well. With a prokaryote, the chance always exists that it will give away its most defining power when it passes a section of its DNA molecule to another. The gene pool of the whole prokaryotic world is unaffected when one bacterium passes a precious genetic segment to another, but the generous bacterium itself suffers a significant loss. With eukaryotic cells, such depleting giveaways are eliminated. Sappho releases sexual cells that are flush with precious genetic information, but Sappho herself remains as genetically potent as ever. The genetic powers that were achieved by billions of years of evolution are maintained in the undivided cell even after it has created new chromosomes for the ongoing adventure.

So too, the differentiation of each Sapphoid line becomes ever more heightened. A proto-sperm cell can enter and combine only with a particular kind of proto-ovum cell. The correspondence between the DNA has to be close, for otherwise dysfunction and death will result. In this way, sexually compatible communities of Sapphoid cells emerge. With these divisions in life's evolution, the particular achievements of an ancestrally related community can be preserved.

With prokaryotes, on the other hand, the law of the lowest common denominator presides. No informational riches can be accumulated in one being, for such riches are equally distributed throughout the kingdom. The prokaryotes can be considered in a sense a single organism with manifold bodies. But with meiotic sex, and the consequent sexually restricted groupings, if one species makes tremendous evolutionary advances along a particular strategy of life, these achievements will be protected by meiotic sexuality's requirement that both sperm and ovum cells come from the species group. In this way, both mating cells will possess the advanced and defining characteristics of the lineage. This power of preserving the achievements of particular lines of creativity led to the fourth and final emergence of the Proterozoic era, the multicellular organism.

THE SURPRISE OF the multicellular animal is not that cells would congregate together. The surprise is not that a group of cells would develop molecular sinews to keep themselves unified in a permanent way. The symbiotic relationships we have already seen would need very little genetic adjustment to stabilize exosomatic connections into the fleshy fibers of a multicellular body, and the particular advantages enjoyed by such a loose congregation are easy to imagine. Even 1.3 billion years ago microphytes—multicellular plants—had emerged almost simultaneously with the development of the eukaryotic cells. It is not multicellularity by itself that is the biological breakthrough. The fourth surprise of the Proterozoic is the emergence of a new and higher-order autopoiesis. Multicellularity enabled new subjects to appear in the universe.

Around seven hundred million years ago, a group of Sappho cells—sexual eukaryotes variously differentiated from each other—formed an alliance. At the beginning, the cells kept their own identities intact. Each was capable of organizing itself with its particular powers. Each had its own orientation in the seas with its own patterned relationships. But in close and perduring associations within a group of intimates, Sapphoid cells would begin to specialize. Those activities at which one cell excelled would be drawn upon more than other activities which could be carried out by another Sappho cell within the association.

The needs and the individual production of each cell would be communicated chemically through the walls held in common by two or more cells. For instance, a surplus of the energetic glucose molecule produced by Sapphina and poured into Sappherella would be interpreted as, "Energy needs satisfied, no need to produce glucose; shut down the appropriate segment of DNA." Thus would molecular messengers in Sappherella make their way through the cytoplasm, settle on the DNA, and stop Sappherella's own production of glucose. So long as some other cell continued supplying Sappherella with such molecules, Sappherella could concentrate its own energies on other tasks, some of which might even be beneficial for nearby Sapphoid cells.

Such communication between two cells or among a particular number of cells must have continued for great stretches of biological creativity. But a moment came when a Sappherella cell received a chemically based electrical message that was not sent by any Sapphina. Nor did it originate in the autopoiesis of Sapphonia, nor from Sapphranida, nor from any other Sapphoid cell. One day some information arrived that was quickly translated into molecular activity, information that altered Sappherella's DNA or RNA or protein production. But even if Sappherella had been able to interrogate all the members of her community she would not have been able to find the cell that sent the message. If she had gone through them cell by cell, each would have denied having created the information.

They would be telling the truth. The sender of the message was not a Sappho cell. The generator of the message was Argos, a new creature. Argos, the new center, the new subject that this community of cells had evoked—it was Argos who sent the message. What had begun millions of years ago as a loose conglomerate of symbionts floating freely in the sun-filled seas had passed a critical threshold and had drawn out from pure potentiality a living, acting being. A new cause had appeared in the world, a power that could not exist but for the intimate communion of those particular cells, but a power that, once allowed to exist, immediately set to the task of ordering the community in new and surprising ways.

Nothing could have been further from the autopoietic awareness of the Sappho cells than the idea of evoking a new being, Argos. So too, when the atoms of hydrogen floated in a vast cloud five billion years ago, nothing could have been further from their patterns of interaction than the details of giving birth to the Sun. In each case a new level of autopoiesis was attained, a new centered and self-governing power entered existence.

The first multicellular animals were as much a surprise as the emergence of the galaxies. For more than three billion years the primary relationships had been among cells and between cells and the inanimate elements. Now Argos appeared with a mind of its own, training ten thousand cells on its own particular aims. Stupendous creativity had been required for the emergence of Argos, and now it fed with ease on individual cells, undoubtedly disrupting a great many ancient communities.

T HE PROTEROZOIC ERA rose out of the destruction of the Archean when overpowering clouds of oxygen smothered the ancient kingdom's vitality. The emergence of Prospero, a being capable of employing oxygen for its life, mutated into Viking's form that burrowed through cell walls and usurped alien DNA for its own purposes. A cooperative alliance involving the aggressive Viking with the long-suffering Engla led to the eon's epochal creativity, that of the eukaryotic cell, Vikengla.

Vikengla mutated into Kronos, who dared to eat living beings, inaugurating an adventure that brought forth ecosystemic communities of predators and prey, meiotic sex with its powers to accelerate evolutionary development, and the one-pointed self-organization of multicellular beings.

In each of the era's four major inventions—the endo-symbiosis of Vikengla, the exo-symbiosis established by Kronos, the sexual-genetic symbiosis of Sappho, the multicellular symbiosis of Argos—life's cosmogenesis advanced. Cells related within cells, cells related within cycles of the ecosystems, and cells related to each other through their branching ancestral trees, added new intimacy to the interrelatedness of the single vast gene-swapping organism of the prokaryotes.

All this proterozoic creativity, which took place almost entirely in the microcosmic realm, ran into a wall. Soon after the emergence of multicellular Argos, the planet became engulfed in possibly its most extensive glaciation in the entire history of life. Even Australia, which hovered then at the Equator, was heavy with ice. A massive extermination of the planet's organisms took place.

An entire era with two billion years of adventure came to its shuddering close, but its biological inventions would weather even this vast glacial extinction and rebound to establish a new episode in Earth's adventure, the Phanerozoic. In the Phanerozoic, life added a distinct layer to the universe; the microcosm and macrocosm gave birth to Earth's mesocosm.

Grizzly Bear, Chinook Salmon, Boreal Forest, Alaska

7.
Plants and Animals

PURE ENERGY EXPANDED, then splintered, into a trillion galaxies whose creativity brought forth at least one planet with a four-fold mode of action: a churning lithosphere, washed by the great gyres of the hydrosphere, evaporating into the billowing winds of the atmosphere, and all folded into the electrified intensities of the biosphere. Earth took its quantum of nuclear energy given by the supernova Tiamat and created a cybernetic system that kept itself in a rich metastability, a great ball dancing at the crest of a spraying fountain of water. In the center of such rich energy, life brought forth the wonders of the eukaryotic cell, meiotic sexuality, ecosystemic communities, and the multicellular beings.

A pervasive desire among early animals is to increase in size. A larger animal can escape the rule of prey. Fewer of the early predators will feed on an animal that has grown beyond the size of their toothless mouths. From the other perspective, a predator who increases in size will be able to draw ever more of the life world into its feeding patterns. When life had, through long trial and error, established a functional Argos—a viable multicellular creature—such animals proliferated rapidly by entering the adventure of exploring the oceans at the end of the Proterozoic era.

The animals grew and diversified to become the early jellyfish, sea pens, and flat worms of six hundred million years ago at the dividing time of the Proterozoic and Phanerozoic eons. All such animals entered the "visible" scale, geologically speaking, overnight, and their expansions in size quickly met the upper limits inherent to cosmological dynamics.

An animal that is too large will be too feeble to live vibrantly in Earth. For instance, a jellyfish is one hundred thousand times—five orders of magnitude—wider than one of its constituent cells, and thus has the size advantages for feeding on unicellular organisms. But if the jellyfish were to expand again two or more orders of magnitude, the electromagnetic bonds holding the cells together would not be strong enough to withstand even slight impacts. Its membranes would tear apart in even gentle ocean waves. Given the autopoietic shape of a jellyfish, its functioning has an optimal size. This optimal size is established partially by Earth in the sense of the dynamics of biosphere and hydrosphere, and partially by the universe in the sense of the strength of the fundamental interactions that were set into a perduring form at earliest symmetry break.

The emerging animals at the start of the Phanerozoic strove to increase in size until they entered the range of their optimal existence. This range of the optimum for multicellular creatures is the *mesocosm,* a domain in between the microcosm of the molecules and the macrocosm of planets and galaxies, a realm with its unique dynamics. If any creature from any phylum of Earth's life were to be expanded toward the dimensions of the Milky Way, or shrunken to the dimensions of helium, the creature would perish far, far before it reached either of those other two realms.

If the Flaring Forth had been even slightly different, life's mesocosm would be vastly different. In that sense, then, the mesocosm can be understood as a specific, quantitatively particular, possibility that had been set into the structures of the universe from the beginning. In examining an animal of the mesocosm one sees a being whose shape has been roughly hewn by the dynamics at work since the origin of time. The invention of the mesocosm is an evocation of an adventure implicit in the earliest fire of the cosmos.

In cosmogenesis each fundamental breakthrough evokes a multiplicity of possibilities. In each new era, creativity explodes to populate the realm with an abundance of novelties, many of which do not survive. In the earliest universe the wild exotic particles were replaced by a calmer set of hadrons and leptons that became the standard particles for the rest of time. When the universe was given the opportunity to fashion galaxies, it brought forth a great display that was then sculpted down as many of the strangest and most exotic galaxies were captured and folded into the standard elliptical form.

When life entered the mesocosm it too exploded with new structures—the structures of advanced animal forms. All future forms of animals are music played on the basic themes established at the crossover from the Proterozoic to the Phanerozoic, six hundred million years ago. Great adventures in differentiation remained, but the most primordial creativity responsible for fashioning the fundamental body plans would soon be finished. The animal phyla that survived the early catastrophe are all Earth would ever see. No new phyla would appear. All future animals would employ the same structures that appeared at that time—as the jellyfish, the annelid worms, the sponges, the starfish, the snails, the sea urchins, the vertebrates, the nautiloids, the insects, the brachiopods, the crustaceans, the spiders, and a dozen or so less prevalent forms of animal life.

THE PHANEROZOIC EON has three eras: the Paleozoic, from 570 to 245 million years ago; the Mesozoic, from 245 to 67 million years ago; and the Cenozoic, from 67 million years ago to the present. Each is characterized by its own biological and geological creativity.

One of the first great inventions of the Paleozoic era was the hard shell. By using such minerals as phosphorus and calcium, previously naked animals could protect themselves with the shells common among trilobites, clams, and snails. Most of the Paleozoic animals fed on the algae of the oceans, but heterotrophy of advanced animals took place within the mesocosm as well. Shells did not eliminate such feeding. Responding to the challenge of such shells, nautiloids developed a beak with the power of snapping through the shell to the fleshy insides of their prey.

Another kind of shell was produced by the early fishes, who developed out of the family of worms. The ostracoderms invented a bony plated armor to protect themselves, and they quickly became a principal predator along the ocean bottom during the middle Paleozoic. Able to wiggle and to glide up from the ocean floor, they fed more effectively than other forms of life along the bottom and initiated a creative experiment that would require two hundred million years to complete. Ostracoderms were quickly crowded out by more effective predator fish. The first major advance was the jaw, enabling a fish to bite down hard. Jawed fish had a much vaster domain in which they could hunt. Later genetic mutations enabled paired fins to be added to the basic jawed fish body, enabling far greater stability in movement. These placoderms soon reached sizes of thirty feet as they soared through Paleozoic seas. A final development—fins supported by fin bones—led to the ray-finned fish, a form of animal so effective in its movement it proliferated through the oceans, eliminating altogether the armored form of fish that had started the era.

The first heroes to venture onto land were plants. The animals dared not. This was not because of any particular difficulty with air. There were most likely already some animals who could handle air with ease. When we reflect upon the nature of the continents back then, we can speculate that to have eyes was to eliminate any courage. From the tip of the sea, one faced a lifeless, thousand-mile expanse of baked rocks and rubble and dust, barren as a moonscape. There was no living soil. There was nothing green. Hell with its truculent citizens would have been more inviting. But the ultimate obstacle to movement onto land was even more ferocious. On land life faced an invisible enemy, a reality so pervasive and so strange and so overpowering that life remained exclusively in the oceans for ninety percent of Earth's history. This invisible power came to be called, hundreds of millions of years later, gravitation.

Gravity is an ordering principle of the universe whose effects are suspended for organisms in the oceans. Thus life learned enough about the electromagnetic powers to create a haven where the fierce demands of gravity could be forgotten, at least for a time; and had it not been for the ancient heroes of the Paleozoic, perhaps for all of Gaian time.

Waves splashed high and smashed across the granite rocks, then slid away in a foaming betrayal of their living cargo. Plants that had mastered the three-dimensional swirl in the ocean currents now found themselves flattened and baked by the sun. How many tons of life perished into crisp green flakes that scattered across the lifeless continents? In the sleepy semi-consciousness of those plants, how many experiences of distress as they lay plastered to the rocks?

Yet the genius of creativity swelled within those plants that were unavoidably engaged with this world-encompassing enemy. There at the edges of sea, continents, and air, a new being appeared, Capaneus, a hero who invaded an alien world. Capaneus invented the wood cell and became the first terrestrial creature able to withstand the flattening power of gravity. Capaneus built these solid structures with vascular vessels to transport food and material through its body. Possibly Capaneus began as a semiaquatic plant surviving at the water's edge, and then developed enough strength to endure gravity as water levels dropped back or the seas dried up. Capaneus's descendents improved their vascular transport and deepened their root system as they formed forests along the edges of the rivers and oceans and throughout swampy areas. These lycopod plants required the wet and dampy areas for their reproduction. Just as with their ancestors in the oceans, lycopod sperm needed a moist world to find their way to lycopod eggs.

The novelty that completed the Paleozoic plant creativity was the naked seed organism, the gymnosperm. Now trees themselves without need of a moist surface could bring the male and female gametes together to create a seed. Having fully mastered the challenges of both gravity and aridity, such gymnosperms marched across the continents to become full forests that redounded back and replaced the previously dominant lycopods.

Just as the ray-finned fish represented the most masterful design for fish life in the ocean and became the principal marine vertebrate, so too the naked seed plants represented the Paleozoic's supreme accomplishment in bringing the mesocosm into the previously uninhabitable dry continental interiors, and so became the principal form of terrestrial plant life.

The last of the Paleozoic's achievements was the mastery by animals of dry land. The first animals to follow Capaneus 425 million years ago were arthropods, probably millipedes, soon to be followed by their predators. These arthropods met the challenge of land by developing an exoskeleton that could keep water within. They became, in a sense, mobile living ponds.

Such insects underwent a profound proliferation as they entered and adapted to the immense number of new worlds found within the lycopod and then gymnosperm forests and swamps. Dragonflies developed wing spans of eighteen inches. Millipedes grew to eight feet in length. One scorpion was large enough to kill and consume small vertebrate animals when they appeared later in the Paleozoic forests. By the last part of the Paleozoic, the insects had achieved subtle adaptations that would never be significantly improved upon, becoming enduring features of the terrestrial biosphere.

Vertebrates joined the adventure onto land 370 million years ago. Some 10 million years previously, at least one branch of the fish family had developed lungs. With so much life having already invaded the land, any fish that could survive even briefly along the shores of the rivers, lakes, ponds, or oceans would have found its way into a paradise of food. Not only were there trillions of insects, the vast majority of these would not be capable of recognizing the large hungry object as a predator. Having evolved in a vertebrate-free forest, the insects would have no instinctual responses stored in their DNA for protecting themselves from such predation. These air-breathing fish would need only the most rudimentary locomotion. Of all the types who attempted this move, the lobed fish proved to have a slight advantage anatomically and thus prevailed as the form of vertebrate to survive on land.

The descendents of the lobed fish, the amphibians, retained the ancient fish strategy of laying eggs in the water. Amphibians began their lives as fish, as tadpoles with gills, growing through their juvenile stages in the water; but then quit the seas, grew lungs, and made their way into the world of swamps and forests, reaching sizes of twenty feet in length.

THE ENTIRE PALEOZOIC ERA, the era that evoked all the phyla of the animal world, that brought forth a succession of fishes and coral reefs hundreds of miles in length, that invented the first forests, and the air-breathing, land-dwelling animals, was completely destroyed 245 million years ago. Climates shifted. Perhaps comets collided with Earth and drastically altered the atmosphere. When the dust settled and the extinction spasm was done, a vastly impoverished animal world was all that remained.

Mass extinctions puncture, in a sporadic manner, the entire fabric of life's story. During the Paleozoic era at least three major catastrophes took place: at the end of the Ordovician; near the end of the Devonian; and at the close of the Permian. This last catastrophe, the Permian-Triassic episode, was the most devastating in the four billion years of life's adventure. In each of these, life in the tropics suffered the most extensively. The coral reefs were extinguished. Much of the warm-water brachiopods and mollusk life was eliminated as well. At least half of the marine families disappeared in the Permian catastrophe, and overall between 75 percent and 95 percent of Earth's species were erased in the destruction.

The biosphere was exceedingly slow to repopulate the animal kingdom. Entire ecological niches, such as those filled by the coral reefs, were left vacuous for millions of years. The new era, the Mesozoic, began with a bleak world compared to the late Paleozoic. This barren world immediately set to the task of regeneration, but the task was exceedingly arduous and time-consuming.

The terminal Permian mass extinction decimated the animal species of both land and sea, thus eliminating vast chambers of genetic information. The few species who survived the extinction were placed in an alien vacuum. Previously they had existed in well established ecological relationships with a variety of other creatures. In such genetically recognizable geological and climatic conditions, their own epigenetic unfolding was assured. But after the Permian extinction spasm, the other species were simply absent, and the planetary conditions were alien. For instance, the great land mass of the planet had come together to form a single supercontinent, Pangaea, which had drifted over the South Pole, creating great glaciers, lowering the seas and bringing forth a colder and drier climate. It was in this alien world, with so much of the former community absent, that the remnant species had to find their way.

The existence of mass extinctions brings us back to our reflections on the violence found in the universe. With the destruction of the vast majority of the species of the Paleozoic era, the opaque dimension of the universe shows its reality once again. Whatever the cause of such extinctions, whether extraterrestrial collisions, or plate tectonics pushing the continent over the pole and bringing about a mass glaciation, or some other major shift in the climates, it is certainly true that an individual species with its particular needs and capabilities is not consulted. The placoderms that had represented a pinnacle of evolutionary design and adaptation were silently erased from the adventure of life by the harsh realities of a suddenly cold, Devonian ocean. In one sense, placoderm learning, placoderm meaning, and placoderm knowledge all vanished as a wisp of smoke.

Mass extinctions rip apart the accomplished interrelatedness of Earth's community. The remnants have genetic memories of the former community of life but find themselves in terrestrial or marine devastation. They cannot rely on what was effective previously, but often must develop new ways of relating to enveloping realities. With the ever-recurring genetic spontaneities, the surviving species have a plethora of new ways of life to explore. The advantage for such species is the very vacuity of the world, which addresses each new anatomical or behavioral invention with the same shrug of the shoulders—"Go ahead, try it out." The tight constraints of a world of rich community, such as the later Paleozoic, were replaced by the "anything goes" world of the decimated Mesozoic era.

INVERTEBRATE LIFE IN the seas slowly recreated itself. Those surviving species radiated and established the new character of Mesozoic oceans. The ammonoids, surviving only as two species, expanded into 150 species. This slow expansion through adaptive radiation was a move typical for the surviving gastropods as well as bivalve mollusks, brachiopods, crinoids, urchins, and sponges. Gone forever were the trilobites, the fusulinid foraminifers, the lacy bryozoans, and the rugose corals. Replacing these were new creatures such as the hexacoral that would eventually join in the reef-building, the coccolithophores, an algae that filled the seas, and the belemnoids, squidlike mollusks without an external shell.

The vertebrate world of the Mesozoic was radically new. Two novelties appeared near the transition from the Paleozoic to the Mesozoic that altered the character of terrestrial animal life. The first crucial invention was the amniotic egg, which freed terrestrial amphibians from their dependence on bodies of water for their mating processes. The egg, protected from the environment by a more or less impervious shell during the period of the embryo's development, is completely self-sufficient, requiring only oxygen from the outer world.

The beings who created the eggs were the first reptiles. They could now deposit their eggs far from the predators who patrolled the shallow waters in search of fertilized eggs. As their skin grew more watertight the reptiles could roam far inland, even into deserts. Amphibians could never reach such regions, for their permeable skin would give way under any prolonged heat, leaving them desiccated. But now the reptiles with their eggs had found a way to continue the adventure into the dry valleys, up into the hot mountains, out into the endless Sun-baked plains.

The reptiles crowded out the amphibians to become the predominant terrestrial vertebrate. Just as the naked seed plants, better adapted to the drier conditions far from the sea, populated the continents and became the principal plant life in the late Paleozoic and early Mesozoic, so too, the reptiles, better adapted to the challenges of the dry climate far from the oceans and lakes, populated the continental land masses, especially after the extinction spasm that inaugurated the Mesozoic.

The second great invention of the terrestrial animal world was endothermy, the power to maintain a warm body even in the face of a cold outer world. The therapsids were the first to enjoy this new capability. Now able to sustain high activity through prolonged periods, even when the Sun was blocked out on a cloudy day, such "warm-blooded reptiles" altered the history of life. The legs of the amphibians and early reptiles came out at an angle, making a crouching posture necessary for their movements. But therapsid speed demanded appropriate anatomical changes and soon their legs were rotated beneath them, enabling a much higher velocity of locomotion. Over the first million years of the Mesozoic, the swift therapsids quickly devoured their way to the top of the food chain, acquiring an even more powerful snapping jaw along the way.

Reptiles also took to the seas. Seal-like placodonts with their shell-crushing teeth discovered their niche when they learned to feed on the rich molluscan life on the sea floor. The nothosaurs also lived along the shorelines and fed in the shallows. The speedy ichthyosaurs competed directly with the fish and sharks when they learned to venture far from land by giving birth not to the amniotic egg but to the live, baby ichthyosaurs.

The plant world was not decimated by the Paleozoic mass extinctions in the way the animal world was. The transformation of the forests from lycopod to naked seed plants continued through the terminal Permian as well as into the Mesozoic. The most common types were the conifers, the cycads, the gingkos, as well as the spore-bearing ferns that proliferated beneath the canopies of these gymnosperm forests.

THE GREATEST MESOZOIC creativity among the terrestrial vertebrate world was the dinosaur, a further development of the warm-blooded reptilian line of therapsids and thecodonts. Dinosaurs ranged in size from a couple of feet to a hundred feet in length. Many of them benefited from the anatomical modifications of their ancestors, and with their quick movements they soon replaced the lizards in many niches. For one hundred million years dinosaurs were the most prevalent vertebrate form. They were social animals that often traveled and hunted in groups. And dinosaurs developed a behavioral novelty unknown in the reptilian world—parental care. Dinosaurs carefully buried their eggs and stayed with the young after they hatched, nurturing them toward independence.

In the middle of the Mesozoic era the first bird appeared, 150 million years ago, a direct descendent of the dinosaur. Birds retained as well the endothermy and the parental care of the dinosaurs. Somewhat later, 125 million years ago, the first marsupial mammal appeared. These mammals shared with the dinosaur such things as vertebrate anatomy, warm-blooded metabolism, and parental care, but differed in having body fur instead of dinosaur scales, giving live birth instead of laying reptilian eggs, and nursing their young, rather than having independent young.

The new mammalian mode of nourishing the young in the earliest period of their existence outside the womb was immensely significant for the future psychological formation of the mammalian species. This bodily intimacy during pregnancy and after birth can be associated with the distinctive emotional qualities that develop in this line of descent. Mammals remained small and probably nocturnal throughout the rest of the Mesozoic.

Other vertebrate novelties of the Mesozoic included the turtles, the crocodiles—both of which reached huge proportions—and the frogs.

The greatest Mesozoic creativity in the plant world was the flower in Cretaceous times. The sexuality of the flowering plants (*angiosperms*) was an order of magnitude more fecund than that of the gymnosperms. Where a conifer would require eighteen months to produce its seeds, a flower could grow from a seed to a mature plant capable of releasing its own seeds, all in a few weeks. Added to its fecundity was the symbiotic relationship between the insect world and the flower. Insects drawn to the nectar unknowingly transport pollen from one flower to the next, fertilizing the plants on which they feed. Often a particular insect will feed only on a particular flower, thus assisting in the process of creating a new species. If a new shape of flower appears, it may draw a different kind of insect, thus isolating this new flower from sexual contact with its original group. In this way new kinds of insects create new kinds of flowers. The flowering plants quickly diversified and spread throughout the continents. Their ecological success depended on such intricate symbioses, in what is sometimes called a survival of the most cooperative.

With the expansion of the flowering world came the proliferation of the insects, and such changes worked themselves back into the vertebrate world with amazing consequences. Dinosaurs had evolved to feed on the gymnosperms, but the flowers were now pushing gymnosperms out. Birds and mammals, on the other hand, fed happily on the flowers and their seeds, and on the symbiotically aligned insects, all of which enabled the birds and mammals to flourish while the dinosaurs declined.

THE MESOZOIC WORLD of sea reptiles and dinosaurs was ripped apart 67 million years ago by what was perhaps the fourth most devastating mass extinction—the terminal Cretaceous. Eliminated forever were the many and diverse forms of dinosaurs, marine reptiles, ammonoids, many rudists, and bivalve mollusks. The coccolithophore algae that had predominated would never again attain such a planetary presence. Earth grew cold and the Cenozoic era began with an impoverished fauna.

In the devastated world of the early Cenozoic, an animal invention of the previous era exploded into many different forms. About 114 million years ago the first placental mammals had appeared. By giving birth to live young, the placental mammal had the advantage of an offspring that began its life at a more advanced form than the offspring of either the reptiles or the dinosaurs. By itself, this invention would not have made much difference in a Mesozoic dominated by the dinosaur, but combined with the mass extinction that depleted the fauna and eliminated the competition, placental mammals increased to become the prominant vertebrate form of the new Cenozoic era.

The Cenozoic is a time of stupendous creativity. Within twelve million years of the Cretaceous mass extinction most of the living orders of mammals—including bats and whales and primates—had already entered existence. This period represents only 2 percent of the mesocosm's history. Thus in a geological instant the placodonts, the nothosaurs, the plesiosaurs, the ichthyosaurs, the mesosaurs, the tyrannosauruses, the brontosauruses, and the triceratops gave way to the Cenozoic's newly created horses, cattle, rabbits, bats, walruses, whales, dolphins, elk, primates, elephants, rodents, and lions.

The plant world once again was relatively unaffected by the mass extinction of animals ending the Mesozoic. The flowering plants continued to diversify, and by thirty-seven million years ago took on the forms that we would recognize as our own present-day flora. With the diversification of the flowering plants there was of course a corresponding increase in the abundance and variety of insects, which enabled predators of these insects, such as frogs and birds, especially songbirds, to multiply. With the growth of frog populations there followed a rapid development of many kinds of snakes.

The continents drifted into each other and brought forth such ranges as the Alps, the Himalaya, and the Rockies. During mid-Cenozoic times continental rifting changed Earth's systems in ways that became decisive for the biosphere as a whole. Antarctica split off from Australia, with profound consequences for the developing Cenozoic forms of life. With Australia moving northward a sea passage was opened up between itself and Antarctica so that the cold currents of water could circulate about Antarctica without being forced toward the equator and into contact with warmer waters. As a consequence, the first sea ice began to form around Antarctica, and waters of freezing temperature sank to the deep sea and traveled northward, surfacing in warmer climates and thus drawing down the temperatures of the planet as a whole.

In a similar way, rifting between Greenland and Scandinavia enabled the previously isolated arctic waters of the North Pole to spill down and affect the climates of Eurasia and North America. Ice caps swelled, temperatures plummeted, climates grew colder, and all the life forms adjusted. For instance, what had been luscious forests now became dry savanna, and one previously tree-dwelling primate group learned to survive in the open plains by walking upright and by bonding together in small family groups. Thus developments leading to the human were initiated at least in part by the continental rifting at the poles.

These shifts in the geophysiology of Earth created a new global climate, an ice age, one with a hundred-thousand-year cycle. After some ninety thousand years of glaciation, the ice retreats for a period of some ten thousand years, and then returns, repeating the pattern. This sequence of glaciation followed by brief interglacial warming has persisted now for three million years, right up to our own time.

Another consequence of the drier climate was the evolution and radiation of the grasses and the herbaceous plants that filled up the plains as the forests shrank. Rodents sprang up and found a new niche in this grassy world, and with their increase came a further development of the snakes, one of the few predators that could follow the rodent into its home. The grassy plains also drew forth an increase in the numbers of grazing animals—the horse, bison, cattle, goat—with a consequent increase and radiation of the predators of such grazers, such as the lion, the cheetah, the dog, the hyena, and the great cats.

AFTER THIS COMPRESSED survey of the creativity of the Phanerozoic we direct our attention to the most fundamental question of all. What is it that shapes life? In our telling of life's story we have moved from single prokaryotic cells to the emergence of complex animals with their great ecosystems of forests or coral reefs. We have moved through the proliferation of fishes and dinosaurs and arthropods. Confronted by the immensity of such biological creativity, we need to reflect on the core of this dynamism, on the organic center of life's adventure. What is it that enables, that shapes, that evokes this awesome ongoing event?

Life's journey is shaped by three fundamentally related, though distinct, causes: genetic mutation, natural selection, and conscious choice (or niche creation). We will consider these in turn.

Genetic mutation refers to spontaneous differentiations taking place at life's root. Though there are various forms of mutations, they all involve the appearance of a newly ordered sequence of genetic materials of the DNA within some living organism. Such mutations of one kind or another occur in an ongoing manner, estimated at the rate of one for every hundred thousand replications.

Genetic mutation is a primal act within the life process, and is fundamental in giving shape to the life story. Mutation is a face ultimacy wears. To ponder its nature is to reflect upon one of the most primordial powers in the universe.

To approach the reality of mutations, we need to approach the reality of *chance, random, stochastic,* and *error.* In wondering over mutations, and in trying to understand them, those who have reflected on their nature most deeply have concluded that they are random. The mutations are an expression of the stochastic nature of life. They are the result of mistakes, of errors. The very powers that emerge in the mutated offspring are brought forth by chance. Each of these refers to this same reality of mutations, but each adds a slightly different nuance.

Random refers to the experience earlier humans had when watching galloping animals. They dart this way and that, they vanish from sight behind the trees, they break across the plains from their hiding places. They run with a fierce freedom. To say that mutations are random is to show respect for the primality of a mutation. A mutation is as unpredictable, as spontaneous, as close to the root reality of the universe as is the quantum surprise of a proton or the sudden spring of a great cat.

Stochastic and *error* are closely related to one another. *Stochastic* reaches back to our early awareness of the pathways that come from guessing, of the directions one takes when following a scent or a hunch only dimly felt as right. *Stochastic* points to the experience of moving toward a goal, but in darkness, and thus with a great deal of groping and multiple approaches. *Error* reaches back to the experience of wandering humans, perhaps lost and confused, moving blindly in a region they know nothing about. *Error* refers to that experience of drifting and changing directions, all in the hope that if one continues to wander, one might stumble upon what is desired. Thus both *stochastic* and *error,* in the context of the differentiations of mutation, refer to an opaque dimension of evolution—a process having no fixed goal, but a process of creativity haunted by a sense of direction, by the vaguest hint of the more fertile way. A mutation is an error in the sense that the mutated form might just lock onto an unforeseen but wonderful moment of creative impulse.

Chance brings to mind the ancient experience of throwing objects into the air and letting them fall to the ground. When the objects left the hand, they left the world of human control and entered the outside, the universe beyond, the realm over the wall of control. Bones slip free of human fingers, of human wills, of human ideas, and enter the preternatural powers governing the universe. Let them show themselves. Let them shape this pattern. Referring to mutation as *chance* brings to mind the whole vast differentiation of all the seas of Earth. Let every form be tried. Throw them all into the air. Let the great dimension of reality participate in working things out at the microcosmic level.

Another word that names this reality variously referred to by chance, random, stochastic, or error is *wild*. A wild animal, especially one on the search, alert and free, moves with a beauty beyond the predictability of a machine, far beyond the lock-step process of a rationally derived conclusion. The *wild* is a great beauty that seethes with intelligence, that is ever surprising and refreshing for the human mind to behold. To say that mutations are a fundamental dynamic in the first cells of Earth is to say that rooted at the core of life is the wild freedom to wander, to grope, to change spontaneously, to run galloping as an animal, even as an animal dazed in search of something. The discovery of mutations is the discovery of an untamed and untamable energy at the organic center of life. Not only is such creativity there, it is centrally there. For without this wild energy, life's journey would have ended long ago.

NATURAL SELECTION IS life's power to sculpt diversity in a creative fashion. It took the genius of Charles Darwin to understand this particular dynamism—to notice it, to appreciate it, to recognize its causal nature. Before Darwin no one in the history of thought had the powers of imagination combined with the detailed observational data necessary to conceive of natural selection. Natural selection, the differential survival of individuals within a particular population, shapes life at a fundamental level.

In a species of woodpeckers in a deciduous forest, there will be a genetic variation leading to a number of different beaks—different in length, width, shape, strength, and so on. One or several of these beaks will be especially effective at prying off loose bark and devouring some particular local insects. This one form is relatively better adapted to the forest than the other forms; other things being equal, the birds happily endowed with such beaks will flourish and produce the greatest number of offspring. Thus if we consider the gene pool of the population as a whole, there will be an increase in the occurrence of those genes that participate in expressing the better-adapted beak. A selection pressure is exerted on the gene pool by pressing toward that particular gene or group of genes enabling such beaks.

Relative to any stable species, the environment has a fixed shape within which the populations are in flux. The genetic variation that sprouts up randomly by way of mutation and sexual recombination produces a spectrum of different responses to the environment. Some birds find a way to make their living within the contours of the environment, and so reproduce and flourish. Others fade away in comparison.

We can think of this dynamic under the metaphor of sculpting. The activity of natural selection sculpts these sprawling genetic variations within the population by carving away those types that are relatively less successful in this particular environment, where *environment* refers to all that affects survival and reproduction—climate and weather, neighboring organisms, geological, chemical, and mineral conditions, inner as well as outer. Natural selection is a survival of the "fittest" in the sense that the genes enabling one particular phenotype to succeed relative to all others are selected and passed on.

Whereas mutation reveals the reality of the random, of chance in the world, natural selection announces necessity. The genes of the bird who cannot secure shelter and food to survive, or who cannot secure a mate, will evaporate at the bird's death. Or if the climate changes, the genes of those birds who cannot adapt to this change will disappear. There is no arguing here, no plea. Only ultimacy. There is the necessity that goes beyond the definitions and reasonings and bafflements of the human mind: life's adventure proceeds with those who survive and reproduce and with no one else.

CHANCE AND NECESSITY are the first two powers that shape life. The third is niche creation, or more generally, conscious choice.

In our interpretation of the evolutionary story of the universe, we have at several points extended beyond the ordinary scientific accounts. We have, for instance, adopted the unitary point of view when discussing the fundamental physical interactions of the universe, taking as our perspective the assumption that the four interactions are representations of a more

primordial universe interaction which through the course of time has constellated into the four fundamental interactions. Even though this point of
view can be said to represent the thinking of a central portion of the theoretical physics community today, it is not yet widely adopted by the scientific enterprise, although we expect that it will become common in the
next century.

Again, in discussing the relationship between the data of our Earth-
based observations and the nature of the evolving cosmos as a whole, we
have extended the Cosmological Principle to include a consideration of the
form-producing powers of the universe, and have organized our interpretations by means of the Cosmogenetic Principle. Even though our knowledge of morphogenesis and cosmogenesis is in its infancy, we are assuming
that the heightened scientific investigation of these dynamics will make the
cosmogenetic perspective entirely ordinary in the next centuries.

So too, here in our discussion of the evolution of life forms, we are
expanding our theoretical framework beyond the ordinary focus on natural
selection and genetic mutation to include a third cause, conscious choice.
This represents no radical break with the neo-Darwinian tradition, for biologists since Darwin have been well aware that conscious choice is causal
in the shaping of life. The question is not whether conscious choice
is a shaping power; the question is whether or not conscious choice deserves to be considered as important as either genetic mutation or natural
selection.

Before we consider the relative importance of conscious choice with
respect to natural selection or genetic mutation, we will consider some ex-
amples to make clear what it is we wish to bring our attention to.

We can imagine a moment when our population of woodpeckers finds
itself at the edge of its world. The valley and its broad leaf maples and
towering beech trees begin to give way to the firs and brush of the mountain biome. What would happen if a population—or more likely a few
birds from the original population—made the decision to hunt the mountains? What if some of them decided to enter a new world to find their
grubs, to mate, to make their nests, and to enjoy in every way the life here
in the world of the mountains?

By doing so they would immediately enter a world with new and different selection pressures. The genes that had formerly been favored, such

as those giving a shorter and more curved wing, would find themselves bypassed in this world. Now the selection pressures would keep those genes that gave a longer and less curved wing, one better adapted for flight in the more sparsely populated mountain forests. The favored beak size and shape would change too; many of the previously favored characteristics would no longer be selected. Different selection pressures would be set to work on genetic variations.

If in a secret viewing room we could magically list the DNA strands of the daring population that had wandered into the mountains, we would see a gradual change in the nature of the gene pool. What is the cause of this change? Mutations? No, for these particular changes would not be random; they would be moving under a particular gradient to the favored set of genes. Natural selection then? In one sense, yes, for the selection dynamics of the mountain would be involved. But it was the decision of those few leaders, to flee the valley and its lowland marsh life and to invade the mountains, that brought about these new selection pressures in the first place. That particular bird's spontaneous decision is the blade that hews the genetic materials of this woodpecker population.

The recognition that animal consciousness has been at the root of biological change also follows from the discovery of organisms whose anatomies do not fit their own habitat as well as they would fit a different habitat. Darwin himself exclaimed over one such example, when he noted that not even a naturalist, when examining the dead body of a water ouzel, would suspect that this member of the thrush family was subaquatic, that it made its living by using its wings underwater. Anatomically, the water ouzel is no different from the rest of the wholly terrestrial thrush family, and yet there it is, making its way in a watery niche. In a similar way the aquatic wasps, who are homologous with their terrestrial cousins, decided at one time to invade the water. And then there is the recent occurrence in the Galapagos Islands of an iguana lizard that has taken to a life underwater to feed on the algae there.

In each of these and related cases, the conscious choice to invade a new niche is the primary step toward subsequent changes in genetic material. It is quite possible that in the distant future none of these above species will seem out of place. The ouzel or the iguana of that future time will differ from today's forms as natural selection pressures sculpt genetic variations to produce a phenotype that blends more harmoniously with the new habitat. Future humans will most likely not find an organism that fits poorly into its world. It is only in this transitional phase before selection pressures have shaped genetic variation that we can see the principal role consciousness has played.

Naturally there will be times when an animal has not consciously chosen to create a new niche, but is simply swept into a new world and makes the best of it. We are not suggesting that the creation of every new niche is the result of conscious choice. We are simply pointing out that in the course of life's evolution, conscious choice has at times been central in shaping the genetic materials, and perhaps the very best example of this, one with both broad and deep implications for the evolutionary story, is that of sexual selection, in particular that dynamic referred to by Darwin as *female choice*.

DARWIN WAS SO certain that female choice was fundamental in the shaping of life's evolution that he refused to be swayed from his position even when confronted by arguments from the co-founder of natural selection, A. R. Wallace. Wallace argued that natural selection accounted for evolutionary modifications, but Darwin maintained to the end that natural selection alone, though it was the primary cause of evolution, could never account for all of life's evolution. If one truly wishes to understand the causes that shape the forms of animals, one must include those decisions that arise in the consciousness of the female when she is responding to the males that come to her and propose themselves as mates. The criteria that she uses, the hunches, the insights, the phylogenetic memories, the mistakes, the sudden impulsive decisions, these are primary causes for the shaping of the face and body of the animal kingdom.

To consider the question of the relative importance of conscious choice when compared to natural selection or genetic mutation, we begin with a quotation from Jacques Monod in which he reflects on profound evolutionary changes and their relationships to the consciousness of animals. "It is evident as well that the initial choice of this or that kind of behavior can often have very long-range consequences. . . . As we all know, the great turning points in evolution have coincided with the invasion of new eco-

131

logical spaces. If terrestrial vertebrates appeared and were able to initiate that wonderful life from which amphibians, reptiles, birds, and mammals later developed, it was originally because a primitive fish 'chose' to do some exploring on land, where it was however ill provided with means for getting about."

Perhaps natural selection and genetic mutation are primary within a relatively fixed environment with relatively stable species. Perhaps in such homeostatic situations conscious choice plays a minor or even a negligible role in life's evolution. Perhaps it is only in the major evolutionary changes, such as the invasion of land, that conscious choice becomes the primary cause explaining the change. At the very least we can say that to understand the powers that shape life, we need to take into account the genetic mutations within such a species of pioneering fish, the selection pressures on that population of fish, and the consciousness of those pioneering fish. To ignore the fish's consciousness is to ignore a real and central cause in the life world. Any story of life's evolution must include mention of that consciousness occurring in the fish.

But why stop with fish consciousness? Mental processes are not confined to the animal world. Though our investigation of the self-organizing dynamics expressed in single-celled organisms has only just begun, we already know that bacteria and protists exhibit behavior that can be interpreted as manifestations of memory, of discernment concerning questions of temperature and nutrient concentration, of a basic irreducible intelligence. During how many moments in four billion years of Earth's history were such minimal powers of discernment called into action? And how many of these primal decisions and how many of these inchoate choices, however crudely carried out, sent the biosphere into pathways forever characterized by those initial decisions?

THE THREE SHAPING powers of mutation, natural selection, and niche creation are further illustrations of the root creativity of the universe we have identified with the Cosmogenetic Principle. Mutation is an illustration of differentiation; conscious choice or niche creation is a biological illustration of autopoiesis; and natural selection is the dynamism of communion.

Random mutations enabled the genome and thus life itself to differentiate itself anew. We spoke of this when we told the story of life's first emergence. The differentiating powers of mutations are a heightening of the differentiating powers leading to minerals or to geological variation. The pressure toward the future within each moment includes a pressure for uniqueness. The universal bias toward the novel, expressed within the life world, is genetic mutation.

Natural selection is a tendency toward interrelatedness as it acts within the ecosphere. Natural selection has been introduced under the model of a fixed environment containing a malleable species, a species that is to be sculpted to fit this environment. By a "fixed environment" we mean the weather and climatic patterns, the other organisms and their patterns of feeding and mating, and the physical surroundings with their patterns, including such things as the chemical changes in the soils and the movements of a river bed. An environment is a biophysical community of beings with a complex pattern of interaction, feedback, growth, and decay, within both seasonal and long-term rhythms.

A species within this biophysical community finds its place and fits into the great complexity, or else the unpleadable necessity eliminates it and all its descendents forever. We can see in this activity the cosmological ordering of communion. Here in the life-and-death realities of the natural world we see the bite of ultimacy where communion is a central reality of life. "Fit into the community and become a fully functioning participant or else you will be left out for good."

But what does this demand consist of? A population of woodpeckers wanders wildly into a mountain community. Everywhere they turn they find demands shouted at them. "Your wings are too stubby; if you want to stay here, fit into our world." "Your beaks are too fat for the crevices in our trees; if you wish to stay here, change yourself." "Your neck musculature is too feeble for our grasses; if you are interested in entering our community, you must pay attention to all of us here and live in this awareness." Each member of the community makes ultimately the same demands on the new population: "If you are to stay, we must become related, and not just externally related. We must become kin, internally related. We live here. Our meaning is here. Our identity comes out of this place of togetherness. If you wish to join us we will work to provide everything you might need.

133

But you must first demonstrate your willingness to let go of your previous accomplishments and enter our world freshly."

We say, "natural selection renders a species well adapted to the environment." This is another way of saying the organism has learned to survive. That is, the community has taken the new species into itself. And why? Because the species has first taken the community into itself. Not just cerebrally, but in every possible way, in a way reflected in its phenotype, in its gene pool. This deep internal relatedness is why nearly every species will perish if plucked up from its world and put down in a different biome. Such creatures do not recognize anyone. The relationships that are worked into their skin and brains are not there to activate them, and so they waste away.

To examine the gene pool and the physical and behavioral patterns of a population of woodpeckers is to see in these the nature of the bark on the community's trees, the nature of the climate and its seasonal changes, the size and chemical composition of the insects, the fabric of the grasses, the shape and habits of the predator birds and mammals. The entire staggering complexity of the mountain community is slowly taken into the heart of the woodpecker's reality, so that its former identity melts away and what lives instead is this mountain-community-as-woodpecker. *Natural selection* is the biological phrase naming this cosmological dynamic of communion. This dynamic of interrelatedness as natural selection is one that presses, always and everywhere, for a deep intimacy of togetherness. This dynamic in living beings goes as deep as the very structure of genes, body, mind; a dynamic that mocks the modern foolishness of thinking that community is composed of isolated atomic individuals.

To be alive means to find one's identity in the togetherness of the community. To be alive means to reveal with varying levels of fulfillment one's surrounding community in every action, feeling, or idea one experiences. Any new population or species that is not engaged to some degree in reinventing itself within the context of its community flirts with extinction.

Natural selection as fundamental to the organic center of Earth's life means that rooted in the core of anything alive is the communal reality. At the heart of the individual is everyone else. There, where one announces to the world who one is; there, where the wild actions burgeon forth; there, in intimacy and togetherness, is the whole web of life.

WHEN WE TURN now to discussing how conscious choice, or niche creation, is an example of the cosmological dynamic of autopoiesis, we need to
expand our definition of environment. For strictly speaking there is
no such thing as a "fixed environment." Climates change. Other species
change. The geochemical conditions change. Everything is changing, individual species and enveloping environment. But as the average pace of
change for a species is usually much faster than the pace of change in
the geochemical surroundings, we can say that the environment is relatively unchanging. Similarly, the pace of change of an invading species is
usually faster than that of the well-adapted resident species, so that the
image of a fixed environment with its changing species assists us in identifying the dynamics of the ecosystem. There is, however, one element that
is entirely missed by thinking of a fixed environment, and unless this in-
adequacy is noted an understanding of the actual dynamics of life's evolution is lost.

Stated rhetorically: a species always creates its own niche. The difference in understanding made possible by this idea can be explained by turning a traditional biological phrase upside down. We say conventionally that
the environment is fixed; within the limits stated above this is a true statement. But it is also true to say that the species itself "fixes" the environment
by choosing one out of a potentially infinite number of niches to inhabit.
We say conventionally that the fixed environment selects the species, but
it is also true to say that it is the species that selects the environment. And
in so doing, the species chooses its own evolutionary pathway.

As mentioned above, this power to choose an environment struck Darwin when he reflected on the water ouzel and its decision to forage underwater. Who can say what will unfurl out of this amazing decision? No one
can. Not even the ouzel. It has simply committed itself to a particular
world, and has thus created a very specialized pattern of selection pressures for itself. Fifty-five million years ago a wild mammal made a similar
decision to taste life in the seas, and out of that decision sperm whales,
belugas, gray whales, humpbacks, dolphins, blue whales, orcas, and fin
whales came forth. Perhaps in the future our human descendents will stag-
ger with wonder at the sight of an ouzel bird-fish, now grown to monumental dimensions. Sweeping silently through the clouds it trains its eyes

on the dark seas below, sky dives, and hacks apart a pod of whales with a beak as rude and competent as executioner's steel.

If so, who created this creature? We want to identify that which is causally efficacious. We want to name the creative and shaping powers of life. Our answer is simply: chance, choice, and necessity. Equivalently: genetic mutation, niche creation, and natural selection. The power shaping life is a wild energy; an inner urgency to pursue a particular path of life; and an immense bonding process that insists upon intimate togetherness. These are the shaping realities of each thing. Nor can any living organism be found that has become what it is outside these powers.

NICHE CREATION, NATURAL SELECTION, and genetic mutation are three faces of an interconnected and ongoing process within the organic world. Genetic variation arises through recombination and mutation, and is sculpted by natural selection; but natural selection itself is constellated and given its form by the particular choice of a niche.

As an illustration, consider the horse and the bison. These hoofed mammals come from a common ancestor, but they are by now vastly different forms of life. What happened? They both occupied the same area, the plains of the North American continent, and both began with similar gene pools. So how is it that they are so different?

One of the primordial ancestors of the bison made a profoundly important choice: when faced with an enemy it would charge head-on. The horse's *ur*-ancestor made a drastically different choice. It would flee all predators. These decisions immediately created two different worlds. From then on, different selection pressures were constellated for each, and these shaped the genetic diversity according to the two different original decisions. To charge and butt creates selection pressures that pluck out such characteristics as thick, squat skulls, massive neck musculature, eyes far removed from the action of the crushing blows. In each generation such individuals are favored with survival and the more adapted form of a bison is thus slowly baked into the genetic powers of the DNA. Similarly, the selection pressures constellated by the horse's decision to flee immediately began sculpting the genetic variation in order to capture a respiratory system that could run for miles, a nervous system that was keenly alert to even the slightest signal of attack, and a skeletal system that could jerk to a full gallop in an instant.

Neither the horse nor the bison had any choice about obeying the necessities imposed by life's pressures. Either they rose to these challenges or they perished. But the amazing fact is that these very necessities were given their ultimate form by the initial choice of niche of the horse and the bison. The world of the bison is in important ways a creation of the bison. Just so, all the difficulties the horse faces are difficulties the horse in some sense set up for itself. Obviously these worlds and these difficulties at one level are not the creation of the bison or the horse. Predators are real. Death is real. The need for water is real. But the subjectivities of bison and horse are also real. Their psychic spontaneities in creating different niches configured their worlds and were thus central to the subsequent genetic shaping of the bison and the horse.

In a biologically meaningful sense, the world that the horse inhabits wears the face of the horse. The horse does not enter a fixed rigid external environment. Instead, the horse's world has taken the horse into account to varying degrees. For instance, the grasses of the plains have evolved in response to the horse's dentition and patterns of grazing. When a horse grazes it countenances in each blade of grass its prior decision to leave browsing in the woodlands and to enter the plains.

The horse always and everywhere works out its relationship with its own macrophase self as manifested in the complex intelligences of its world. To be a horse means to take on the primordially shaped challenges that were set into the world by the horse ancestors. "Flee predators" and "graze the plains" were chosen as the horse's central life decisions. In those decisions the horse began its sacred journey into its destiny. Those choices, constellated in the teeth of natural selection, govern the history of horse evolution.

The choice of the primogenial bison governs its species' own development. The choices of the primogenial squirrel, the pursuits of the primogenial whales, the distinct paths of the primogenial angiosperms, the dream visions of the primogenial wolves, all create different worlds and different pressures that eventuate in the forms of life about us today. Can we say that the primogenial bat chose to become what it is today? Can we hold that the primogenial horse chose to become the great stallions millions of years in its future? Certainly not. Even if these early Cenozoic mammals had had the power of imagination necessary to conceive of future forms of their descendents, they would in no way have been able to orchestrate life to get there. Their first step would already change the world that changes them.

Their movement into their future evolution began with commitment to a vision—a vision strongly felt but seen as if fleetingly and in darkness. Perhaps it was just the sheer thrill of the gallop that captivated the first horse's consciousness and convinced it to make that species-determining decision: "We will run come what may." No vision of itself in the future, and yet the future pressed into its experience of the moment: "Here is a way to live. Here is a path worth risking everything for."

Once the horse committed itself to this behavior, the world in which it lived was transformed. All the elements proclaimed: "You want to become a galloping energy? You may, but only if you include all of us and all of our concerns and realities in your life project. If you insist on becoming the animal who runs the plains, we will teach you what is required. Listen to us and you will gallop. Ignore us and you will starve and vanish." The modern horse is not simply a separate organism. The horse is the whole region galloping, a creation of the entire great plains. A vision of the future pressed into the horse's awareness and it responded, "Why not? Let's see how we would gallop here."

PERHAPS THE GREATEST gift Darwin gave to humanity was the opportunity to see in all of life an ongoing, intelligent, creative drama. Rather than thinking of a form of life as having been put on Earth in a fixed form at the beginning of time, we now see each form of life arising out of the Great Adventure. The ouzel tasted something in the aquatic life that drew it in, that convinced it to pursue that world. So too, the primeval whale when it was still landbound. The great ocean waves curled under the spindrift, crashed into sea foam and promised something. Promised something irresistible, food yes, and another day under the sun, yes, but also life itself, life and its glorious adventure.

BY THE END of the Phanerozoic eon, all five of the kingdoms of life were represented in Earth's community—bacteria, eukaryotes, plants, fungi, and animals. We close this chapter with a sweeping sketch of all five king-doms, the achievement of four billion years of adventure.

The first kingdom, the monera or bacteria, consists possibly of millions of different species of bacteria. At the present time around five thousand species have been identified. These were the first forms of life. They are the hardiest creatures anywhere on Earth. They can live in boiling water. They can be frozen like rocks and then come back to life. They are found higher and deeper and in colder and hotter regions than any other sort of living being on Earth. They are the advanced guard of life. They are always the first to enter an area. There is no community of life anywhere on Earth that lacks monera; indeed they are irreplaceable. A spoonful of soil has an es-timated fifty billion bacteria in it. They are the core form of life in Earth's community.

The next kingdom is the protists or the eukaryotic cells with sixty-five thousand species identified. The eukaryotic cell emerged in the Protero-zoic eon and is the basis of all advanced forms of life. Protists can be divid-ed into three basic categories: the algae such as the phytoplankton, which are thought to be the protoplants; the protozoa such as amoeba, which are thought to be the protoanimals; and the slime molds, considered to be the protofungi.

The third, fourth, and fifth kingdoms—the fungi, plants, and ani-mals—all emerged during the Phanerozoic eon. A hundred thousand spe-cies of fungi have been identified. These are not plants because they do not harvest the sunlight, but rather gather their food from their surroundings by absorbing nutrients through their cell walls. Fungi are involved in the work of decomposition throughout the biosphere.

There are an estimated three hundred thousand species of plants, the fourth great kingdom. The majority of these are the flowering types, an-giosperms. Flowering plants consist of at least a quarter of a million spe-cies, stretching from magnolia trees to orchids to blackberry vines. The remaining plants are the naked seed trees, the gymnosperms, and the non-vascular plants such as mosses and liverworts.

The fifth kingdom is that of the animals, multicellular heterotrophs with developed capacities for digestion. The largest subgroup of those classified is that of the insects, some eight hundred and fifty thousand spe-cies. There are at least five hundred thousand species of round worms and forty thousand vertebrate species. Among the vertebrates there are nine thousand bird species, six thousand reptile species, and four thousand, five hundred mammalian species.

We have classified on the order of two million species of life. Biologists estimate there may be ten to thirty million species altogether. This number represents only one percent of the species that have come into existence since the beginning of life. Billions of life's species have emerged and gone extinct. But though so many forms of life have vanished, there was never a time in four billion years of Earth's life with as many species as there were when the human first arose in Earth's community. Great chasms of experience had been irrevocably lost through mass extinctions, but a beauty endured through it all. The catastrophe at the end of the Mesozoic was overcome by life's fecundity, and the overall richness of life on the planet surpassed that of any previous era. Perhaps the only word to describe the world that gave birth to the human form of life is paradise.

✶ ✶ ✶

Black Bull, Lascaux Cave Painting,
Dordogne, France, c. 15,000 B.C.E.

8.
Human Emergence

FIFTEEN BILLION YEARS ago, shape-shifting matter appeared as primeval fire, to transform into galaxies with their stars and gaseous clouds, then to take on the form of molten planets and to shift again and wear the face of the squirrel and the mosquito and the incandescent root hairs of the towering sequoia and all the billion living species of Earth's adventure over the last four billion years. When shape-shifting matter suddenly appeared in human form a great surprise took place. For a new faculty of understanding was making its appearance, a mode of consciousness characterized by its sense of wonder and celebration as well as by its ability to refashion and use parts of its exterior environment as instruments in achieving its own ends. The story of the human is the story of the emergence and development of this self-awareness and its role within the universe drama.

In its beginnings, and in its early development, the human was so frail, so unimpressive, a creature hardly worth the attention of the other animals in the forest. But these early humans were on a path that would in time explode with unexpectedly significant new power, a power of consciousness whereby Earth, and the universe as a whole, turned back and reflected on itself. Earth's community of life would never be the same, for here was a development with such consequences that it can only be compared to the emergence of oxygen within Earth's early communities of life, a development that carried both the destructive and the creative significance of that earlier event.

THE DRAMA OF self-consciousness takes place in five phases: the primordial emergence of the human; the Neolithic settlements; the classical civilizations; the rise of nations; and the Ecozoic era. The story begins with the appearance of the first primates of the mammalian family of life, some seventy million years ago.

The primate story as we now know it begins in the Cretaceous period at the end of the Mesozoic, not long before the collapse of this period and the rise of the Cenozoic. The primates, including the lemurs, tarsiers, monkeys, apes, and humans, were for the most part forest-dwellers, living extensively in trees. While only a single family of primates was present sixty-five to fifty-five million years ago in the first phase of the Cenozoic, the Paleocene, a more expansive primate development took place in the lengthy period of the Eocene, some fifty-five to thirty-six million years ago. By the end of the Eocene a large number of primates were present in their prosimian form throughout the entire area that includes what is now the Eurasian continent.

In the next geological period, the Miocene, the story moves to Africa, where some thirty million years ago in the Fayum region of the Upper Nile Valley a fox-sized primate appeared, a primate with an anatomy and teeth such as would later characterize the fully developed apes. Until this period Africa had not participated in primate development. From here on, however, Africa became the main focus of developments that would culminate in the earliest hominids leading on to the modern human.

Somewhat later on an island in Lake Victoria in east-central Africa, an early fully formed ape appeared with a brain capacity of slightly over 150 cubic centimeters, generally referred to now as "Proconsul." In this predecessor of the later apes the tail is no longer part of the anatomy. Such was the beginning of the early expansion of primates in Africa, where the greater anthropoids appear somewhat later: the gibbon and orangutan, now native to the East Indies; the gorilla and the chimpanzee, native to Africa; and the hominids. In the line of developments leading to the hominids (also known as the australopithecines) the gibbon separated out some twenty million years ago; the gorilla, nine million years ago; then the chimpanzee and the hominids, some four million years ago. The chimpanzee is closest to the hominids in this line of development.

Many of the most significant developments in this transition period just prior to the appearance of the human occurred in a region that includes Kenya, Tanzania, and Ethiopia. This is the region where some of the earliest prehumans, the hominids, lived some four million years ago. The hominids were distinguished by their increased brain size and by their capacity for walking in an upright posture, although they still spent time in trees, as is indicated by the length and muscular quality of their arms relative to their shorter legs. They lived now mostly in the savannas, the grasslands of the region.

Such was the situation some four million years ago when a young female hominid, now designated as "Lucy," lived in southern Ethiopia. Her brain capacity, between four hundred and five hundred cubic centimeters, was slightly greater than that of the chimpanzee. In diet, Lucy and the other hominids apparently were vegetarians; neither Lucy nor any of the others left any evidence of implements for hunting or dissecting, or of bones discarded after eating.

That hominids were present at this early period is evident also in over sixty footprints left by two hominids walking erect who crossed over a region of volcanic ash at Laetoli in northern Tanzania. These footprints made some four million years ago survive until the present as the ashes hardened after the impressions had been made.

Forest life seems to have advanced the psychic capacity of the earlier primates, the capacity to swing easily among the tree branches, the quickness of mind, and the focusing of attention—all qualities that were preserved when their hominid descendents moved out of the forest to the more open savanna areas.

THE MOST ELEMENTAL physical transformations leading to human identity can be indicated as increase in brain size, upright posture, bipedal walking, frontal focus of eyes and countenance, development of the arm and hand in relation to the eye, increased capacity of the hands for grasping, and incidental use of nature-shaped stones as implements.

The relation of these physical developments to the inner psychic development of the human can be appreciated if we consider that upright posture freed the arms and hands from the burdensome role they had in walking on all four limbs. The hands could develop their prehensile role. This activity of the hand could in turn respond to the eyes more quickly. One important consequence of the use of the hands in grasping objects was to relieve the jaws of this task. This led to a refinement of all those aspects of the throat, tongue, teeth, and lips that would later be used in speech. Then came two most significant developments: enlargement of the brain and the prolonged period of childhood prior to maturity.

In this sequence of developments a change occurred in anatomical structure, especially in the expansion of the cranial vault enclosing a larger brain size, in leg formation, and in the vertebrate column, all leaving the hands free not only for shaping and handling tools, throwing, hunting, and food gathering, but also for artistic expression and emotional communication.

SOME 2.6 MILLION years ago, at the close of the Pliocene period, the earliest expression of the human appears in its species identity, a form of the human designated as Homo habilis. A mutation had taken place in the gene structure. A species transition had occurred. The first human appeared. While the more adequate expression of this mutation would require the long centuries of history for its elaboration, the transition itself was manifested by an increase in brain size and by the new human mode of acting, first through the use of tools, later through the control of fire, and the entire course of human physical, social, and cultural development.

Although the earlier hominids had achieved upright stance, a slight increase in brain size, and occasional use of tools, there was no systematically developed stone industry indicating a fully reasoned understanding of what they were doing. Members of Homo habilis, with a brain size of seven hundred cubic centimeters, were physically more refined in features, with smaller teeth than their predecessors, and more competent in their interactions with the environment. We do not know the full story of Homo habilis, as this first expression of the human is designated, yet we do know that members of this species lived at Turkana in northwest Kenya some 2.6 million years ago.

Of the inner thought and cultural life of this earliest expression of the human we know only fragmentary details from the physical remains left by chance from the daily life of that primordial period. Habilis made an abundance of tools in the region of Olduvai, a place in the Great Rift Valley of Kenya. Some were made of cores with sharpened edges and others from the flakes struck from the cores. These stone implements were used extensively for the capture and dissection of animals.

That Homo habilis were hunters and meat-eaters is evident from the type of implements they made, as well as from the teeth structure and the manner of their wear. This marks a decisive change from humans as exclusively gatherers to humans as both gatherers and hunters. Hunting brought about extensive changes, not only in life-style but throughout the entire psychic functioning of this early form of the human. A new sense of self-identity occurred along with a sense of having a power of life and death over other forms of life much closer to themselves than plants.

The dramatic dimension of existence was considerably enhanced. New skills were developed, new sensitivities, a new understanding of the animal world—their habits of grazing, their movements, their capacities for seeing, hearing, and perceiving scent, their patterns of flight from danger. The earlier knowledge of plants available for food, even the intimacy that developed in relation to the plant world, was considerably enhanced by the new mystique of the predator-prey relationship that developed between the humans and the animals. For humans to kill required something more than the physical vigor and cunning associated with the stalking process: it required a psychic permission granted by the cosmological order itself.

With Homo habilis an event of singular importance takes place: the beginning of the Stone Ages in the cultural development of the human. The shaping and use of stone was one of earliest achievements of humans in their level of thought development and in their distinctive relationship with their environment. Even now the various ways of shaping stone identified as "Paleolithic" or "Neolithic" provides our customary terminology for designating the most significant cultural phases through which these earliest humans passed prior to entering the urban civilizational phase of these past five thousand years, when there was a turn from stone to metals in the shaping of primary implements for so many basic life usages.

Already these early humans possessed remarkable skills in their working with stone, skills that manifest not only a sense of utility but also an emerging feeling for esthetic beauty, for proportion of form as well as for efficacy in function. Exactly here in these transition years the more significant foundations for the human mode of being were established. The sense of time and space was developing; imagination was receiving the impress

147

of its most powerful images; the stock of primordial memories that would influence all future generations was being developed; an intimate rapport between the human and the natural world was being established, a rapport that was filled with both the terror and the attraction of the surrounding wilderness. The ever-recurring sequence of seasonal decline and renewal was making its impress on the human psyche as one of the most basic patterns that would later find expression in ritual celebration.

AFTER HOMO HABILIS, the next phase of human development occurred some 1.5 million years ago with the appearance throughout Africa of Homo erectus. This newly arrived human with a brain capacity of over one thousand cubic centimeters flourished throughout Africa, although our most extensive evidences are from the Turkana region of Africa, the same region where many of their habilis predecessors had lived.

Homo erectus carried out the first grand migration of peoples by moving out of Africa, first throughout the Asian world and then throughout southern and western Europe. Known as Trinil humans in Java and as Sinanthropus in China, Homo erectus inhabited the East Asian and Southeast Asian worlds for many centuries. During this period Sinanthropus lived, often in caves, in the region of Choukoutien somewhat north of Beijing, beginning around 500,000 B.C.E., a period that coincided with the second (Mindel-Riss) interglacial period.

Among the stones used in this phase of Homo erectus in China and elsewhere in Asia, Africa, and Europe were quartz, flint, occasionally obsidian, or even schist, granite, or quartzite. The dominant tool in the earliest period consisted of core stones, often large pebbles taken from streams, with flakes removed, cores that would fit easily into the human hand. Other tools include choppers, scrapers, awls, and chisels, as well as the famous hand axes that play such an extensive role in this Lower Paleolithic phase of human development. During this period animal skins began to be used for clothing. Shelters were made of stakes set into the ground in ordered patterns and covered with branches and possibly at times with prepared skins.

The controlled use of fire at this time left extensive remains of ashes, charcoal, burnt bones, and charred clays found not only at Choukoutien but throughout other sites in Africa and Eurasia. Fire became especially important as Homo erectus moved farther into the Northern Hemisphere both in the European and the Asian worlds, where survival depended on newly invented modes of protection against the cold through clothing, shelter, and the warmth of fire. The controlled use of fire is the first extensive control of the human over a powerful natural force with almost unlimited possibilities that would be associated with human development over the centuries. Together with the shaping of wooden and stone implements, fire becomes the primordial humanly controlled technology.

Beyond its use for warmth and possibly for food preparation, a psychic advance accompanied the physical skill, since fire makes a unique impress in human consciousness. Control of fire gave to the human a sense of power as well as an increased sense of human identity in distinction from other living beings. With the fire in the hearth a communing with mythic powers takes place, social unity is experienced, a context for reflection on the awesome aspects of existence is established. The hearth becomes the place where the basic social unities, the family, the band, the clan, the tribe, all achieve the intimacy needed for effective social cooperation. Around the hearth is security from possible predators as well as the assurance that comes from community bonding.

THE PERIOD COVERED by Homo habilis and Homo erectus from 2.6 million years ago until 120,000 years ago, a period of over 2 million years, can be identified as the Lower Paleolithic period in human cultural development. The deepest foundations of human consciousness as well as the most elementary technologies of human survival were considerably advanced at this time. Already during the third interglacial period in the European region the Abbevillian and Clactonian cultures that developed prior to the hand axe period had culminated in the refinements of the Acheulian. These were all periods of vital cultural creativity as well of technological advancement.

Such creativity was all the more significant since these developments took place in very small groups of hunting peoples living extensively in encampments with some semipermanent cave shelters, with only chance associations with each other. These groups were self-sustaining bands of varying size but generally of some fifteen to thirty persons hunting and gathering and leaving for later discovery only a scattering of fossilized human bones, teeth, the bones of animals killed for food, ashes, charred pieces of wood, pollen, occasionally physical impression in some enduring matter, but yet an immense number of stone implements—the most enduring remains throughout this period, coextensive, with few exceptions, with human presence throughout the various continents.

Among the variety of implements made at this time were points for piercing, primitive axes for different uses, from cutting trees to hunting animals, choppers for cutting into slain animals or other materials, scrapers for preparing skins for use as clothing or for shelters, and stone flakes used as blades.

The finest creation of this period was the hand axe, which became a central identifying feature of this phase of the Paleolithic. It was a defensive weapon as well as a main implement for hunting. Extensively identified with the Acheulian culture of these times, the hand axe and its associated culture was spread throughout a great part of southwestern Europe and extending as far as England in the west, throughout India to the east, throughout Africa as far as the southern cape. It was generally missing in the East Asian regions. At one time, it has been estimated, this hand axe culture extended over half of the human-occupied regions of the world and over one-fifth of the land surface of the planet.

AFTER THESE CENTURIES of development through the period of Homo habilis and Homo erectus, a decisive new development occurred some two hundred thousand years ago in north-central and east Africa: the appearance of Homo sapiens, the human species that succeeded these other forms and from which all contemporary humans are descended. A second great wandering of peoples occurred at this time, a wandering that by forty thousand years ago brought the human beyond the African-Eurasian world to both North and South America, also into Australia. These epic journeys were carried out with amazing genius for finding their way and adapting to the most varied geographical terrain and environmental conditions.

Homo sapiens went through two main phases in its development throughout these centuries: an archaic Homo sapiens phase and a modern

Homo sapiens phase. The archaic phase as this existed in Europe is associated with the Middle Paleolithic culture, which can be dated from
120,000 years ago until 40,000 years ago. The modern phase is associated
with the Upper Paleolithic culture, which began some 40,000 years ago
and lasted until the beginning of the Neolithic period, some 12,000 years
ago.

In the Middle Paleolithic period, from 120,000 until 40,000 years ago,
archaic Homo sapiens occupied a wide area of Europe from Czechoslova-
kia in the east to Germany along the Rhine in the west. In this period these
incoming peoples began using cave dwellings in the hills along the Nean-
der river valley, a district where Düsseldorf in Germany is now located.
From these settlements along the river they came to be designated as
Neanderthals.

From the third, the Riss-Wurm, interglacial period and for some years
afterward these archaic humans dominated the human venture throughout
these regions and then disappeared, leaving little influence as regards physical descent but giving extensive information on our cultural development.

Typical of the Neanderthals was an extensive industry of refined tri-
angular points made of stone flakes. That Neanderthals used fire extensively is evidenced from the ash remains and the hearth sites. Their food,
besides that obtained by gathering fruit, berries, roots, and nuts, was a diet
of reindeer, bison, deer, gazelle, goat, and bird, with fish added where this
was available. Pollen remains and imprints of plant forms preserved in
hardened earth have lasted through the centuries to tell us of the plants
used for food.

Neanderthals were physically robust in size, with a brain capacity of
around 1,500 cubic centimeters, a capacity larger even than the average for
modern humans, which is around 1,350 cubic centimeters. The cranial
bone structure was decidedly more massive than at present, with a pro-
truding face, powerful jawbones, and heavy ridges above the eyes.

The cultural expression of the Neanderthals is generally referred to as
"Mousterian," a name taken from the region in southern France where extensive remains of this culture were found. This culture is identified with
its cave dwellings, extensive use of fire, personal ornaments, and well-
wrought stone implements made from flakes taken from a core piece and
then shaped for more specialized work, especially for scraping and cutting.

Neanderthals also used the Levalloisian technique of shaping stone, using a prepared stone core with a leveled top so that a number of slivers to be used as blades could be struck lengthwise from the top. This was the beginning of shaping blades of some length. Altogether the Neanderthals had over fifty different types of stone implements. Also among their basic achievements was the making of skin-covered huts much more advanced than those of any preceding peoples.

ONE OF THE most significant aspects of the Neanderthal culture is the bear cult that originated this early in human history, a surviving cult that is still practiced by the Ainu of northern Japan. In this Middle Paleolithic period the Neanderthals manifest the beginnings of a religious mode of consciousness, with ceremonial practices associating belief and life meaning with elaborate burial rituals. At one site in Lebanon a ritual burial took place with a deer slaughtered as food for the deceased and placed within a pattern of stones painted with ocher. The consciousness of death is immensely important in the psychic functioning of these early times.

Such a burial site with its ordered arrangement of stones is a mythic, ritual mode of expression that from its beginning is the manner in which humans respond to the universe, just as birds respond by flying or as fish respond by swimming. It is an interpretation of existence consequent on the awesome experience that humans have in witnessing the coming to be and the passing away of things. There appears to be a psychic tendency in humans consciously to integrate their own human process with this cosmological process. From these earliest times humans experienced themselves within the encompassing order of the universe.

This order of the universe was experienced especially in the sequence of the seasons, the dissolution of winter and the renewal of springtime. These were determining experiences of humans from a very early period. This cosmological process observed so powerfully in its seasonal sequence has been an ultimate referent in terms of reality and value. Since it was a never-ending process without historical beginning or ending, to be integral with this process was to attain fulfillment of existence.

This entire range of natural phenomena impinged on human consciousness at this time with a wonder that easily turned into ritual celebration.

The transition moments of the cosmological order evoked awe and reverence and invited participation. That all this was related to danger, to the struggle for survival, to death provided the challenge and excitement that is itself, perhaps, an imperative deep within the entire cosmological process. The universe was a dramatic reality, filled with powers and voices, constituting the Great Conversation that humans participated in through daily and seasonal rituals as well as through rituals associated with birth, maturity, and death.

Here we might observe that the capacities for tool making, finding shelter, identifying food sources, and evoking and using fire were never sufficient for human survival. Associated with these the human needed a rapport with the all-pervasive spirit powers manifested throughout the phenomenal world. Such powers from this early period provided the psychic support needed by the human mode of consciousness. To obtain this support, to evoke that numinous presence perceived as the origin, support, and final destiny of all that exists, humans from earliest times engaged in symbolic rituals, often with sacrificial aspects. Differences in these rituals and their associated activities have inspired the more creative powers of the human intelligence and imagination that have given each of these various cultural traditions their distinctive qualities.

After establishing such a powerful regional presence throughout the Eurasian world and moving the human sequence through a significant phase of its development some forty thousand years ago the Neanderthals disappeared from the scene. There was apparently a diminishment in numbers through natural causes and then an absorption of surviving members into the modern Homo sapiens, known also in Europe as the Cro-Magnon peoples, who had recently moved into this territory.

WHILE THE DECLINE of the Neanderthals was occurring and modern Homo sapiens, the Cro-Magnon people, were moving into the European region, a significant change of climate was also occurring. The last advance of the Ice Ages was in process throughout the Northern Hemisphere, the glaciation period known in America as the Wisconsin glaciation, in Europe as the Wurm glaciation. This glaciation, which began some seventy thousand years ago, had reached its farthest southern extension some eighteen thousand years ago, when a recession began that has continued until the present.

The biosystems of the planet, in both the plant and the animal worlds, were also in a state of change throughout the northern regions of the Americas and the Eurasian continents. Many of the tree and other plant and animal species retreated south to await the opportunity for returning to more northern regions as the glacial chill diminished and the newly shaped contours of the land became visible. At this time too the megaforms were disappearing, among them the woolly mastodon and the mammoth, assisted by extensive hunting by humans both in Europe and on the North American continent. It was also a geologically active period, with a new series of volcanic eruptions throughout the planet.

Immediately upon their arrival in the European region the Cro-Magnon peoples manifested an artistic and inventive genius completely missing in what we know of earlier peoples. Both in the volume of their technological inventions and in their artistic skills the Cro-Magnon were an overwhelming presence on the European scene. This can be observed in over two hundred caves with these people's wall paintings and engravings, while their individual sculptures and engravings might be in the range of ten thousand or more. Two periods in this Upper Paleolithic period were especially abundant in their productions; the Aurignacian (from 34,000 to 30,000 years ago), and the Magdalenian (18,000 to 11,000 years ago).

These earlier centuries of the Upper Paleolithic were characterized by the shaping of stone blades of greater length and more skillful crafting than those of prior times. Through careful retouching of the edges with pressure flaking, often by use of antler or bone instruments, longer and more sharply edged tools could be shaped. More than one hundred different types of stone implements were fashioned in this period. Also many implements and ornamental objects were made of bone or antler or ivory. Some of these were decorated with incised patterns of naturalistic life forms. Among the most impressive of these are the finely decorated spear throwers, implements invented some twenty thousand years ago. Harpoons were made for use in hunting land animals. Lamps were used extensively, with the fat of animals supplying the fuel. There were also fireplace cooking utensils along with a variety of containers, some of them of wood.

The paintings and images of this Upper Paleolithic period have extensive representations of animals of various species: bison, cattle, reindeer, horses, and goats. There are also a large number of representations of humans engaged in hunting animals or, apparently, as shamanic personalities performing ceremonies. In the Gravettian period (thirty thousand to twenty two thousand years ago), an extensive number of small "Venus" figures were sculpted in various materials such as ivory or naturally tinted steatite, generally with special emphasis on the reproductive organs, apparently as fertility concerns. One with special artistic grandeur is the small *Venus of Lespugne,* sculpted in ivory from a mammoth tusk.

Although there was extensive dwelling in caves there was also extensive habitation constructed as pits dug into the earth, covered with plant materials and paved with smoothed cobblestone taken from a flowing stream. At times the stones were heated to sink them into the frozen earth. At least one dwelling above ground was made of the massive bones of the woolly mastodon.

Some twenty thousand years ago the tailored clothing worn at this time in Russia was copied in a small sculpted figure that has come down to the present to tell us something of the life-style of these peoples. Also a man fully clothed in a tailored garment was buried in the Moscow region during these years. The garment has decayed but the beaded decorations remain intact, giving us some sense of the refinements in attire that existed in these years. Superbly wrought needles for sewing were made in abundance. Along with clothing from animal skins various ornaments were fashioned, mainly pendants and necklaces using animal claws, teeth, and bones, as well as shells and colorful stones.

Associated with such craft skills and artistic expression is a parallel competence in technological processes. Considerable equipment as well as advanced skills were required to execute such elaborate wall paintings in the dark recesses of the caves of southwestern France and Spain. There was need of light, of coloring materials, of implements for applying the paint, of scaffolding. It also required drafting skills in designing the images, each in itself and then in relation to the others in the more extensive group paintings. All this involved social organization; these elaborate compositions were produced not spontaneously by unskilled persons, but for social purposes by professionals.

In the animal figures painted or sculpted we can observe a certain naturalist style. Yet precisely here we find expressed a mythic communion and a fascination with the animal world through these magnificent forms. To come upon these by the dim flickering light of small lamps that were available at this time was to enter into a world of mysterious powers. The caves evoked their own mystique that deepened the daylight experience of the outside world. That such forms should also be sculpted on spear throwers and other implements indicates the presence of esthetic feelings in these centuries, as well as a high level of rapport with these all-pervasive spirit forces experienced by the entire society.

SINCE THESE ARTISTIC abilities were associated with a new capacity for the understanding and use of spoken language, we are meeting here not simply with another change in methods of stone working or the processing of some physical material but with a transformation of human consciousness on a scale and with a dramatic impact such as we seldom encounter in this narrative of our emergent human development.

Even in the earlier primordial periods of Homo habilis and Homo erectus some capacity for communication through symbolic modes of expression in sound and gesture must have been manifested, since we observe such capacities even in the animal world. At the human level, understanding, communication, and artistic expression in their distinctive human modes were emerging quietly over periods of time when change proceeded at a pace much slower than we have experienced in recent centuries. Yet the pace was quickening throughout the Upper Paleolithic period.

Not only were human capacities for understanding and communication expanding their range of expression, but so too were the human emotional and feeling capacities. These qualities for intimate association with others, for sympathetic response to the needs of others, is shown especially in the child-parent relation as these exist in the mammalian world, particularly with the maternal parent, since the survival of the offspring depends so directly on personal intimacy between mother and child. So too there is the more extended family and group allegiance, since survival is constantly at issue in these small groups of humans who have only themselves to depend on. This gave rise to that intense feeling for intimacy with the group characteristic of such small survival bands.

In hunting societies the band necessarily remained small, since the game present in any given region was limited in relation to the number of humans who must survive by hunting. Hunting of course could seldom be an exclusive mode of survival. Other methods of foraging were needed, including the gathering of fruits and berries and nuts and roots and even bark. Survival was less an individual or personal thing than a group achievement. Considering that each band was responsible not only for its food and shelter but for all its other needs, for its hunting implements as well as for the hunting itself and for processing and sharing the catch, we can begin to appreciate the range of skills and the amount of traditional knowledge needed by the individual as well as by the group.

Extensive geological as well as geographical knowledge was accumulated as each band became familiar with its territory and searched out the proper quality of stone for its needed implements. Also a considerable amount of biological knowledge resulted from acquaintance with the habits of the animals sought. Anatomical knowledge was acquired from the dissection that took place in the processing of animals for food. Also with regard to the plant foods that were collected, an extensive amount of botanical knowledge was accumulated that was handed down through succeeding generations, just as were the skills for working stone.

A MONG THE DISTINCTIVE aspects of the human from its earliest appearance has been its lack of specialized functioning, specializations that are advantageous in their immediate efficacy but that are also limiting in a larger perspective. Such specialization is found in the insects with their intricate patterns of behavior that are so marvelous and yet that also constitute a fixation. So the birds with their flying, the fish with their swimming, horses with running, trees with blossoming. While each of these has its own grandeur, each is locked within its own perfections. For that which bestows the perfection in that very act imposes the limitations.

To have fins or wings is not to have hands with the immense range of possibilities open to such an organ, which can embrace or repel, bring life or death, build a sand castle or a cathedral, paint the Sistine chapel or play a Mozart concerto. Such an unlimited range of possibilities resides in hands that are freely directed by the human mode of intelligence.

Even so there has been a price to be paid for this liberating mode of the human: loss of the instinctive, almost infallible, guidance that is available to the nonhuman species in their various activities, the guidance that enables the birds without teaching to obtain their food, enact their mating patterns, build their nests, care for their young, sing their songs, and migrate over vast distances. With some species of animals a certain amount of teaching and example is needed. They must learn some of their skills of survival. Yet this learning is minimal and of another order than that of humans.

Because they did not have beaks or horns or fangs or claws or the capacity to hide, because they could not run or fly or pick up the scent of an enemy, humans were vulnerable. Yet because there was bare skin lacking fur to keep warm in winter, it was possible to invent an unlimited variety of clothing and ornamentation for warmth and for the enhancement of appearance. Because there was such physical helplessness during the prolonged years of infancy and childhood, it was possible for extensive mental and emotional development to take place. Such a period was needed to activate both the physical and the psychic capacities. At birth the human was only imperfectly human.

A long acculturation period was needed for arriving at a truly human maturity, for the learning of language, for initiation into the rituals whereby humans coordinate their own activities within the cosmological order, for adjusting to appropriate roles within the social order, for acquiring artistic skills, for learning the stories, the poetry, the songs and music of the community. If these activities were not fully developed in the earliest period of Homo habilis, they were within the human process from an early period. For these acquisitions appear in an advanced form as soon as we discover their earliest expression.

While every life form functions through some inner organizing and directing principle that guides the growing process from its origin in simple cell division on through a sequence of developmental processes to maturity with minimal if any teaching or acculturation, the human depends much more extensively on acculturation processes that take place after birth. Whereas the nonhuman life forms receive their guidance almost completely through their genetic coding, the human is genetically coded toward further transgenetic cultural codings, which in their specific forms are invented by human communities themselves in the various modes of expression.

These distinctive cultural inventions are not simply the work of individuals or even of a single generation. They are the work of a people over the generations. This cultural coding of the human is mandated by the human genetic coding in its basic direction but not in its particular identifying features, as for instance, thinking. Humans are genetically mandated to think and have no choice to think or not to think, although they do have the choice of how and what to think and the uses to which they apply their thinking. Once the process is begun it is handed on through succeeding generations and receives its elaboration through these successor generations.

When we consider the historic task of these earlier human societies to attain a new level of creativity through their reasoned control over their lives, their thinking, and their environment, we can appreciate the time and the genius required for this achievement. To abandon the security of genetic guidance for the uncertainties and the probing processes of reason was dangerous both for the human and for the planet, since reason could bring about ruin as well as splendor on a planetary scale. Yet the freedom of human intelligence opened the way for those highly differentiated cultural inventions that function in the human order as differentiation into distinctive species functions in the nonhuman order.

At the close of the Upper Paleolithic period, some eleven thousand years ago, the great achievements of the Cro-Magnon peoples are suddenly terminated. A discontinuity appears in our historical record. The next phase of the human venture, transition to the Neolithic village, would have its centers more in the southwestern corner of the Asian world and in the northeastern corner of the African world, while new and more distant centers would arise in the far continents of the Western Hemisphere.

The Paleolithic phase of the human had fulfilled its destiny. The human was now fully accomplished in all its basic skills. The various continents had been occupied. The human population of the planet had risen to over a million. The different ethnic groups were settled in their homelands. The basic arts had been created. The technologies for survival had been invented. The caves of the brilliant Magdalenian art would now be closed at their entrances by natural processes and would disappear from human consciousness until suddenly they would be rediscovered some eleven thousand years later, in 1895 C.E., by four young boys playing in a cave near Les Eyzies in the south of France.

* * *

Bird Goddess, Pre-Dynastic Egypt, c. 4,000 B.C.E.

9.
Neolithic Village

T HE EARLIER PALEOLITHIC period when humans survived by hunting and gathering culminated, in Europe, in the grandeur of its Magdalenian phase, with its implements shaped in wood, stone, horn, and bone; its necklaces and pendants for personal adornments; its shamanic capacity for communing with the spirit powers beyond the world of appearances; its esthetic sensitivities and artistic skills as manifested in its cave paintings. All this was associated with a deep indwelling of the human within the natural rhythms of the cosmological order. The primal languages of the various peoples were enabling them to communicate more intimately among themselves and with future generations. Clearly articulated traditions concerning the universe and the role of the human in the universe were being established.

This is the context in which the Neolithic phase of the human species emerged in the long sequence of its transformations, only now at a vastly increased rate of change. After 8000 B.C.E. small settlements were growing beyond temporary encampments or cave dwellings into sustained forms of village life, consisting at first of a few hundred and later of several thousand people. The main achievements of this five-thousand-year period between 8000 and 3000 B.C.E. were the domestication of plants and animals, technological inventions such as pottery and weaving, more clearly articulated social structures, primary architectural structures, house-building and the first shrines, a vast extension in the range of language and expression of thought and feeling in an oral literature and in the visual and performing arts, all this along with an elaborate ritual rapport with the cosmological order and the mythic powers of the universe. The Neolithic village based

163

on horticulture and the domestication of animals was the integrating context of all these developments.

Domestication of plants and animals and establishment of village life came as a vast tide of change that swept over the various human communities in the early Neolithic period and included every continent. Not every human group was included in this change but an overwhelming number of peoples experienced a profound alteration of their human situation. If the first distinctive human relation with the planet Earth was the inauguration of tool-using processes, the second was the invention of horticulture and the domestication of animals within the context of the Neolithic village. These village settlements have been the setting in which most of the peoples of Earth have lived until recent times, even though the great urban centers were established as the focal points of the classical civilizations.

Most of the plants and animals that have served humans in this role were domesticated during the Neolithic period. Although this period varied in different regions of the Earth, the beginning on every continent except Australia had occurred it seems, by 5000 B.C.E. A profound and abiding change throughout the entire structure and functioning of human societies went with the change in relations with the other vital components of the planet Earth. Permanent village communities became possible as soon as horticulture and inter-village trade had developed.

The active intrusion by humans into the functioning of the life forces of the Earth until this time had been minimal, limited mostly in the effects that hunting was having on the animal world. Now there was a more active involvement with the vegetative world first through selection of plants for adoption and then through the breaking of the soil and the planting processes with a digging stick and then the hoe and eventually with the plow. The natural irrigation processes along the river valleys also provided a context for sowing seeds from a former harvest. In all of this there was the bringing of the vital growing and reproductive processes and the energy available from the nonhuman world into the realm of human benefit.

If we look at the rather long period of 30,000 years of the Upper Paleozoic we can see that this Neolithic domestication process was accomplished within a limited time period and in regions as separated from each other as the American continents are from the Eurasian world. There would seem to be some cultural rhythm experienced throughout the human process, a rhythm with minimal or no direct influence from any universal originating center, a rhythm that shows up again in the middle of the first millennium B.C.E. when so many of the basic intellectual-spiritual traditions were articulated in such highly differentiated patterns but functioning in the same basic role of establishing the various human orientations toward the world of meaning and the mysteries of existence in a world of birth and death in an ever precarious context.

Although the significance of the Neolithic settlements extended beyond agriculture, this agricultural basis was all important. A number of regions of special significance could be identified. In southwest Asia and southeastern Europe, there were Neolithic developments in the region extending from the Persian Gulf north-westward through Mesopotamia and Syria into Anatolia, then across to Thessaly in the Greek peninsula, up to Armenia and on into southeastern Europe, a region associated in a special manner with the cultivation of wheat and barley. A series of independent developments occurred in Southeast Asia, where the Neolithic village appears quite early, beginning with rice cultivation and domestication of the pig and chickens as early as 7000 B.C.E. In the Americas the Neolithic emerges in the same basic time period independently from anything happening in the Eurasian world. Sub-Saharan Africa had extensive Neolithic village development from 5000 B.C.E. The Bushmen of South Africa remained wandering hunters and gatherers, although the other tribal peoples of sub-Saharan Africa were beginning a development that would lead to the kingdoms that arose throughout this region in precolonial centuries of our era. The Australian aborigines remained in their Paleolithic phase of hunting and gathering until they were brought into contact with colonial powers from the European world.

As regards domesticated animals, the dog has a special place as the first among all those associated with human communities for the past ten thousand years. Dogs were useful for hunting and guarding and at times for food. The well-developed capacity of dogs for smell and hearing make the dog especially beneficial for humans, whose sight is highly developed but whose sensitivities to scent and sound cannot match that of the dog and so many other animals. The dog is so universally associated with the human that from the earliest Neolithic times the dog is found throughout Eurasia and Africa as well as the Americas.

After the dog the next domestication appears to have been of animals such as sheep, goats, and pigs, that could serve as food, then cattle and reindeer, animals for transport such as the camel and the horse, and the water buffalo and the ox for cultivation of the land. In South America the alpaca was domesticated for its wool and the llama for transport. In the Eurasian-African world sheep were also herded for their wool.

An immediate consequence of this new mode of attaining food by planting and harvesting wheat and barley, rice and millet and yams, squash, beans, corn and potatoes, was a more settled existence in permanent shelters clustered in village communities. The difficulty of soil exhaustion was not resolved but alleviated by the natural processes of periodic flooding of lowlands, by field rotation, irrigation, and by clearing the land of trees by burning and cutting. Fertilization by organic materials was discovered early since domesticated animals existed along with plant cultivation from the earliest Neolithic times. The earliest instrument for cultivating the land was the digging stick. Then came the hoe. These two instruments dominated the horticultural process of the Neolithic villages.

Pressures of food supply for a rising population played a significant role in leading the human occupants of a region to begin the seasonal cycle of sowing and reaping, thus increasing the yield of what previously had been gathered from the wild predecessors of these plant forms. This increase in food supply in turn brought about a further increase in population. This necessitated migration of significant groups of inhabitants to other regions. From this early period, even from the Paleolithic period of hunting and gathering, population has been a primary factor in the survival and development of human communities.

A MONG THE FIRST clearly identified Neolithic villages of this early period in the Middle East, we find Jericho in the region of Palestine along the river Jordan, Jarmo and Hassuna by the upper Tigris, and Çatal Hüyük in central Anatolia. A shrine to the mother-goddess located by an oasis in the Jordan region in the late Paleolithic centuries had become a place of gathering, which, around 8000 B.C.E., led to a more permanent clustering of dwellings in the region that came to be known as Jericho. The huts built here of tamped earth and Sun-dried brick were sufficient in number to indicate that a true village was in existence in these early centuries. As soon as the settlement achieved some permanence, a further need for instruments suited for a peasant horticultural economy inspired the shaping of stone sickles for cutting the grain, baskets for harvesting, pots for storage and as water containers, stone querns for grinding the grain.

Hunting continued, since transition to a village context did not involve the sudden abandonment of earlier ways of obtaining food through gathering, hunting, and fishing. Yet the tendency everywhere was toward diminished dependence on hunting and gathering, with ever-increasing reliance on plant cultivation and, in some places, increased emphasis on pasturing animals.

Jericho, with its three thousand inhabitants at this early period, can be considered as an archetypal form of the numerous villages that came into being in these centuries when a decisive revision was being made in the relation of the human species with the natural world. The ancient Paleolithic identity of the human with the functioning of the natural world and the entire cosmological order was changing into a more conscious understanding and adjustment to the natural systems whereby humans themselves had come into being. Humans began responding to the natural world in ways that would be increasingly beneficial to the human in terms of shelter from the elements, a more regular supply of food, clothing better suited for the environment, and more systematic rituals whereby the human would experience its new role in the universe.

The basic discovery in this new situation was that the reproductive power of plants could be enhanced by human care that the seeds of the more useful plants be placed in nourishing soils, be tended, and that the harvest be divided between what was used for food and what was needed for sowing at the proper season in the following year. To carry out this program required an understanding of the orderly sequence of nature through its seasonal cycles. Although this sense of rhythmic order in the seasonal sequence was as ancient as the human itself, its precise calculation was never so necessary as when horticulture came into being. The more effectively humans entered into the natural sequence through horticulture, the greater their need for understanding its functioning. The calendar eventually became one of the main instruments of the civilizational process.

Beyond the pragmatic purposes of horticulture, however, there was from the beginning a mystique of participation in the cosmological order of the universe thought of not as mechanism, but as a vast dramatic enterprise of interacting spirit forces, all expressions of the abiding presence of the Great Mother deity. The drama of the universe was primarily expressed in the seasonal sequence of autumn decline and springtime renewal. Participation in this process brought about the inventions of both horticulture and the domestication of animals. Humans were expected not simply to take from nature but to enter into the productive processes of nature. The animals also belong primarily to the Great Mother. Not only were they cared for as sources of food, they were also brought into the domestication process through association with the birth-death ritual in sacrifice or in some other manner. For while in its earliest form the veneration of the Great Maternal Presence took place through the offering of fruits and flowers, the very symbolism of the seasons and the violence inherent in the natural processes led in time to the extravagances associated with orgiastic rituals.

Shortly after the founding of Jericho in the Jordan valley, the settlement of Jarmo appeared on the north bank of the upper Tigris river some hundred and fifty miles north of Baghdad. The ancient sites of Hassuna and Halaf in this same upper region of Mesopotamia go back to this same period, while Nineveh emerged shortly afterward. All of these sites are far to the northwest of the delta region where the later civilizational developments took place in such places as Susa, Ubaid, Ur, Erech, Lagash, and Nippur.

Of these early beginnings Jarmo is distinguished for establishing a horticultural settlement dating from the pre-pottery period. At Jarmo are found all the basic implements of the early horticulturalists, the polished ax and adz, sickles, the quern, dishes shaped of polished stone. We find here also the bone remains of animals already or about to be domesticated: the sheep, goat, ox, and pig. Although the capacity for fired pottery was missing, we do find painted pottery, the first such pottery so far discovered. Clay statuettes of the mother-goddess so widely worshiped in the late Paleolithic and throughout the Neolithic period are present here in abundance.

At Hassuna along the west bank of the Tigris a more advanced mode of fired pottery, painted in linear or geometric patterns, was developed, indicating an artistic spirit far removed from the naturalism expressed in the animal figures of the late Paleolithic period. In their horticulture, their pastoral activities, their housing, and their furniture, as well as in their skills for designing and firing pottery, the inhabitants of Hassuna represent an integration of the emerging Neolithic culture in one of its finest expressions. These achievements can be dated at around 5500 B.C.E. The influence of Hassuna extended throughout this region of the Middle East.

Another Neolithic settlement of major significance at this period can be identified as Çatal Hüyük in the south-central region of Anatolia—a region more commonly known geographically as Asia Minor, in modern political terms as Turkey. Here a Neolithic settlement existed as early as 7000 B.C.E. In its full extent covering an area of some thirty acres, Çatal Hüyük contains some of the most brilliant achievements found anywhere in this period between 8000 and 3000 B.C.E., the five thousand years that carried the human venture from its late Paleolithic period through to the archaic civilizations of Mesopotamia, Egypt, Crete, and the Indus Valley. Çatal Hüyük is associated with a larger area of Neolithic development that extended to the north and west from the Ukraine through ancient Thessaly and the Danubian to the Alpine regions of southeastern Europe.

The Çatal Hüyük community cultivated wheat, barley, and other legumes along with all the domesticated species found in the more easterly regions of Palestine and Mesopotamia. The pottery is the most advanced of its period. Not only were polished stone implements and vessels produced, but copper and gold were also fashioned. That sailing vessels were known can be seen from their depiction on the pottery.

Many of the best qualities of the human found their expression here, where the goddess was venerated with special care.

Within the Near East, the first spread of wheat horticulture from its early development in the region from Anatolia to the Persian Gulf was to the Nile valley. The Nile valley during these centuries was a contained place bordered on both east and west by desert, a thin green line of vegetation extending for some hundreds of miles along the lower reaches of the Earth's longest river. Significant geographical differences are found in these two regions and in their cultural expression. The Middle East is broken up in its contours with its diverse river systems, its upland regions, and its deserts. Also the weather conditions, especially the coming of the rains, are less predictable in their timing and more violent in their nature.

Egypt in its soils, its vegetation, and its wildlife was immensely more fertile in these early Neolithic centuries than in later times when changes in climate, erosion, and predation on its wildlife caused the region to become more austere in its demands on the human population. The original shaping of the Egyptian mentality was influenced from this early period toward a more even sense of the cosmological order than in Sumer, where the weather patterns were much less predictable. In Egypt we have a less dramatic sequence of natural phenomena. Even though the inundations of the Nile valley were so extensive, they were more regular. Life was more stable. There was a feeling for eternity.

That the Neolithic in Egypt was closely associated with that of Mesopotamia is especially clear in its grain cultivation, since the wild ancestors of its wheat strains are found not in Egypt but in the hills in the upper regions of the Tigris-Euphrates basin. The earliest settlements in Egypt are found in the region between Merimde in the Delta region and Badari in the upper reaches of the river. Much of the earliest development was in the Fayum region with its large lake. They very quickly adopted the basic horticulture of the Mesopotamia region.

M OVING EASTWARD FROM the Iranian plateau there was a culture drift across central Eurasia of the basic Neolithic ideas, as regards both grain cultivation and the arts associated with the making of pottery. Although the climate, the terrain, the native vegetation, and the animals differ considerably from those of the more westerly regions, basic influences from the Middle East pervaded this area, as can be seen clearly in the Neolithic phase of pottery scattered throughout this region.

In its southern regions, with their rice cultivation, China is more related to southeast Asia, while in its northern regions, China is more closely related to the developments across the Pamir mountains and Afghanistan in Mesopotamia and Anatolia. There were rice gardeners with water buffalo in South China by 7000 B.C.E. By 5500 B.C.E. there were millet farmers in North China, where the traditional Chinese civilization had its origins in the eastern regions along the Yellow River. Influence from the Near East in these early times passed across the Pamir and Hindu Kush mountains, then either through the foothills of the T'ien Shan mountains to the north or along the Kun Lun mountains to the south to the Tunhuang region, then down the Kansu corridor into central China, a route that later became the Old Silk Road from Chang-an to Iran and on to the European world. In this region of western China at Lin-hsi and Ulan Hada we find some of the earliest sites of Neolithic culture, although it needs to be remembered that in these centuries the peoples occupying this region were not related to the historical Chinese.

Influence in this region from Mesopotamia and Anatolia is clear in the pottery designs, especially in the painted patterns of the Yang-Shao settlement. Just as the polished black pottery developed at Çatal Hüyük came after its early painted pottery, so we observe a similar sequence here. That this influence continued on across China to the Shantung peninsula is clear from the polished black pottery found there at a slightly later period.

Thus northern China, in its early Neolithic phase, forms the easternmost boundary of a cultural region identified as the Painted Pottery Culture extending from Anatolia, the Black Sea and Thessaly, through Mesopotamia and Persia to Baluchistan in the southeast, and in the northeast on to Chinese Turkistan and as far as Anyang and the Shantung peninsula. This Painted Pottery is identified by its spiral and meandering decorative patterns as well as the numerous similarities in form, especially the tripod with hollow legs. What is more impressive is that the earliest productions of this pottery in the various regions are the most impressive in their esthetic qualities. The later productions are qualitatively less refined in both their form and their decoration.

AMONG THE IMPRESSIVE Neolithic cultures of southeastern Europe and the Aegean, Crete was among the most spectacular and the longest lasting. A long Neolithic period had taken place in southeastern Europe and the Aegean by the end of the fifth millennium B.C.E. when a more advanced stage of the Neolithic appeared and a fully developed civilization emerged shortly afterward. These early developments led to one of the most exquisite civilizations developed anywhere, the Minoan civilization that emerged here in its full expression by the third millennium B.C.E.

Neolithic settlements existed along the Black Sea, in ancient Thessaly in the Greek peninsula, in Italy, throughout the Danube valley, across central Europe, and into the Rhineland. Western European settlements, including those in Britain, can all be dated prior to 3000 B.C. Yet until this time western Europe was culturally not an advanced region. In eastern and southeastern Europe, in Thessaly, in Rumania and Transylvania, and in the lower Danube advanced Neolithic cultures did exist. With the coming of the Bronze Age the Mycenaean and the Celtic cultures both advanced across Europe. They were predominantly Indo-European chieftain societies committed to the warrior ideal. They worshiped a sky deity, although they also preserved some symbols and goddess names from the culturally advanced agricultural settlements they came to dominate. Even from this early period western Europe was a patchwork of peoples and cultural developments rather than a unity, a situation determined largely by the geographical configuration of the region with its divisions into well-defined sections such as the extensions into Spain and Italy to the south, Switzerland in the Alps, the central plains, the northern lowlands, the offshore islands of England and Ireland.

While western Europe was entering its mature Neolithic period, after 2000 B.C.E. the civilizations in Mesopotamia and Egypt were already far advanced in their technological achievements, especially in their entry into the Bronze Age after 3000 B.C.E. The ceramic and visual arts were far advanced. Writing and record keeping had gone on for centuries. There were temples and shrines of vast dimensions requiring extensive bureaucratic organization of the societies as well as mathematical knowledge, engineering skills, and architectural design of a high order. All this had taken place while western Europe was only attaining its first horticulturally based village economy. This "barbaric" aspect of western Europe originated in and was periodically reinforced by the incoming over the centuries of warrior societies from the East.

Shortly after 2000 B.C.E. a unique development took place in the Megalithic structures found from Sicily in the Mediterranean along the Atlantic coast as far as Britain and into the Nordic regions of Europe. These structures in the late Neolithic period were generally formed of immense stones used as sides and as covering for a tomb or shrine enclosure apparently in service of religious ritual directed toward the Great Goddess. Sometimes, however, a single great stone, called a Menhir, is erected, or a series of such stones placed in a circle or in a row. Since these monuments are erected in barren or desolate areas they evoke an overwhelming response at some awesome, some mysterious presence. The feeling expressed by these monuments is that felt before the mystery of death considered as a return to the womb of life whence we emerge at our birth. These monuments may also be integrated with a solar observance, as is possible at Stonehenge or in some other manner associated with the vast forces at work in the cosmological order.

I N CENTRAL AND South America a Neolithic culture developed in a radical independence from these developments in the distant Eurasian world, yet so remarkably coordinated in the time of its development and in such features as its pyramids, its great altars, and its megalithic images, with the larger patterns of development in the Eurasian-North African continents, that we are attracted toward acceptance of greater cultural connectedness than we can presently account for, whether by some historical contact with a continuing immigration from East Asia or by some inner psychic similarities moving to a unified rhythm in their historical expression.

In the period from around 6500 until 3500 B.C.E. inhabitants of the Tehuacan Valley region of what is now the state of Puebla in Mexico developed a form of horticulture that included squash, peppers, beans, and an early variety of corn. Associated with these settlements were querns for grinding the corn, bowls shaped from stone, and permanent shelters. Pottery appears only around 2300 B.C.E., derivative it seems from the more advanced pottery from the northernmost regions of South America, Colombia and Ecuador, where pottery was already being made by 3000 B.C.E. At this time also villagers along the Pacific coast of Guatemala lived extensively from the foods available from the sea in the form of turtles, crabs, clams, and a wide variety of fish. If we date the rise of a truly advanced civilization such as the Olmec on the Gulf coast to 1150 B.C.E. then this interval between 6500 and 1150 B.C.E. is the time when the Neolithic transition occurred in this region.

173

To the south along the Andes the earliest cultivation took place before 3000 B.C.E. Gourds, squash, cotton, amaranth and quinoa were grown. The more extensive development occurred along the coast, where there were numerous villages with several hundred inhabitants, possibly as many as fifty thousand in total population. Pottery was being made here by 2300 B.C.E. Fabrics were also being made, in an advanced twining process that made it possible to color the strands and then to weave them in such a way that representations of various animals as well as humans would appear. Prior to 1000 B.C.E. the heddle-loom had been invented and used to make cotton fabrics. Peanuts originating in the Amazonian region were being grown.

While numerous edible plants were cultivated in the Western Hemisphere, only a few animals were domesticated. There were in the Americas no wild predecessors of sheep, goats, pigs, horses, cattle, or chickens. In South America we do find the alpaca cultivated for its wool and the llama for transportation. As in other parts of the world the dog, descended apparently from the gray wolf, was domesticated and found associated with Indian tribes at whatever level of their cultural development.

The supreme horticultural achievement in America was the development of corn. The genetic difference between the wild species from which it is derived and the cultivated species as this was developed through the centuries is much greater than the genetic difference between the wild and the cultivated species of wheat, indicating an extensive skill in bringing this plant into its role as a principal sustaining food for the peoples of the Americas. This crop, originally developed in the region of what is now Mexico, spread quickly, along with squash and a variety of beans, throughout the temperate region of North America as well as into South America.

WHILE WE CAN identify the time period when the Neolithic village came into being, we cannot identify any terminal phase, since the Neolithic village as a relatively self-sustaining community has been the enduring context for the human mode of being for the past ten thousand years. The various urban civilizations can themselves be considered as elaborations of the basic Neolithic inventions. Developments in language, religion, cosmology, the various arts, music, and dance all occur in their primordial modes of expression in this period. If the Paleolithic is the dawn of the human, the Neolithic is the early morning when the human awakened to a new range of its powers. A case could be made for keeping the village as the normal context for human development, with the provision that there also be a limited number of larger urban centers where some aspects of the human might find more extensive forms of expression. There does at least seem to be a case for establishing a more effective rapport between these two types of community that enable the human to function in its full efficacy. The difficulty is that the large urban centers have assumed the role of being the ideal context for developing an authentic mode of the human, and that village life is considered essentially a retarded life and an unworthy context for the truly human.

In reality the Neolithic village has been the exciting context of immense creativity. It was as peaceful a period as humans would ever know. In Neolithic Old Europe, for instance, we find little evidence of warfare instruments or fortifications prior to the Bronze Age; walls and weapons were the exception rather than the rule. In the social order it was a period of community intimacy and apparently the widespread sharing in decision making. It inaugurated new artistic motifs.

In most Neolithic villages women were at their high moment in providing cultural, religious, and social leadership since identity at this time was generally given in terms of matrilinear descent. Three of the great inventions on which the Neolithic village was founded were most likely discovered by women: horticulture, pottery making, and weaving. The domestication of animals might also be due to women providing the leftovers from their harvesting for the more responsive antecedents of domesticated sheep, goats, cows, and pigs. In the arts also, in decoration of pottery, in clothing design, in personal ornamentation, in all these aspects of human life, women were clearly present in a primary role.

Beyond all these observations we might indicate that the universe itself was experienced primarily in its inexhaustible fecundity and thus primarily identified through the image of woman. The fecundity of the universe identifies with the deepest energy of the universe as its primary mode of expression. That this was the basic cosmology of the Neolithic village is witnessed by the thousands of female images discovered throughout the Neolithic world. The intimate association of the peaceful goddess with the early earth-based religions ritually integrated with the seasonal rise and decline of the vegetative cycle has been so enduring that the Neolithic village continued this primordial devotion as a subculture, even after the warring male sky deities took over control of the various societies as a ruling class with an entirely different ethos from that of the peoples over whom they ruled.

Both the dominant cultures and the subcultures were integrated with the cosmological cycles. In the one case, however, the cosmological order was seen in its hierarchical dimensions with a supreme ruling authority communicating its absolute power to a human ruler. In the other case the ruling principle of the universe was seen as a shared power dispersed throughout the universe which formed a comprehensive sacred community. That this mode of thinking should be so compatible with the horticultural process and with a dominant goddess worship seems clear enough. The later transition from female to male deity, from Goddess to God, made when the Neolithic village was superseded by the rising urban centers, can now be recognized as one of the more portentous transitions that have taken place in human affairs.

With the indigenous peoples of North America we find a naturalistic all-pervasive sense of Mother Earth, with no clear reference to divinity or goddess status. This special way of relating to the Earth in its maternal aspects goes with the profound commitment to an all-pervasive Great Spirit referred to by various names as Orenda, the Manitou, or Wakan Tanka, experienced also as Grandfather. This ultimate principle of the sacred as creator, as providential guide of all aspects of the natural world, goes also with the sense of the human as integral with the natural world.

If the sense of divinity as feminine is especially clear in the earliest phases of the Neolithic and in its later most brilliant expression on the island of Crete in the Minoan culture of the third millennium B.C.E., the more cosmological dimension would be expressed in the earliest forms of the Taoist traditions in China, with their sense of the primacy of the soft over the hard, the yielding over the assertive, the origin of all things in a mothering principle.

AMONG THE MOST difficult to deal with, if also among the most significant, aspects of the Neolithic period has to do with language. Here we have the human in its most intimate mode of self-expression. By language the human articulates its state of conscious awareness within and gives expression to this conscious awareness without. The intellectual capacities of the human begin to function at a new level of efficacy. Reasoned thinking can develop. Communities of thought can be established. By language humans achieve a greater intimacy with each other. The sharing of thought takes place. Education has new dimensions of efficacy. Advanced social order becomes possible. Enduring traditions are established.

In the later Paleolithic period thousands of languages found their incipient modes of expression. In the earliest phases of the Neolithic these languages were brought to a new richness of expression and thousands more were invented. We know of no human societies without language, nor do we know any language that does not have a unique power of its own possessed by no other language. Nor do we know of any language that has not undergone extensive adaptation over the centuries. Nor are there any languages incapable of interpreting to some degree the thought expressed in other languages, since the language of the mind is not in any particular language.

There are indeed interrelations of languages, even families of languages intimately related to each other, such as the Indo-European in Eurasia or the Algonquin languages in America, but so far we have no way of establishing any encompassing unity in the historical derivation of the various languages. The languages of the world remain among the most astounding monuments to the creativity of the local communities that sprang up throughout the various continents. One language that comes down to us from this period in a great number of different forms is the Indo-Aryan, or Indo-European, language. Much more than a language or group of languages, the Indo-European is a vast cultural movement that has taken control over a large area of human consciousness and has possibly influenced if not controlled a larger range of human affairs than any other language. In its Sanskrit form the Indo-European became the basic scriptural language of Hinduism and the language of the basic Mahayana texts of Buddhism. This was in addition to the variety of forms it assumed in the European world.

What can be said is that this period of the Neolithic did, in general, establish more of the power words in the languages of the planet than any other period. We will never know vast numbers of these Neolithic-derived languages, since so many of these have disappeared over the centuries, especially in more recent times. This linguistic loss represents the loss of cultural variety and the rise of monocultural and monolinguistic regions that are an immense and irretrievable cultural impoverishment. Even the languages that survive tend to lose contact with these primordial power-words, words that took on their form and meaning at that moment of total intimacy of humans with the natural world and with the deepest immersion of the human in the mysteries of existence itself.

This was a revelatory moment in the most significant sense of the term. The rituals whereby humans enter into communion with the cosmological order originated at this time in the late Paleolithic and the emergent Neolithic. The archetypal symbols communicated to the human in its genetic endowment were activated for the first time in this setting. It was a magic moment, a moment of psychic innocence that would never occur again, a moment that we return to constantly in our efforts to understand the true meaning of the words that we use, words that determine our most profound sense of reality and value.

It was a moment when the human was able to establish its species identity with a new clarity, an achievement that had its admirable but also its dangerous aspects since this clarity of species identity tended toward isolating the human within itself over against the nonhuman components of the larger Earth community. Once again we can observe that every perfection imposes limitations. Liberation in one aspect implies bonding in another. With language the larger dimensions of the human community become possible; the city emerges on the horizon. The difficulty has been to keep the plasticity of language, to keep language from fixation or trivialization. The inner dynamism of the Neolithic village, its every achievement, was pushing the languages relentlessly on toward further transformation.

When thinking about the Neolithic we might consider its first creative phase as existing from 8000 B.C.E. until 3000 B.C.E. when Sumer, Egypt, and then the other archaic civilizations arose.

A second phase would be the continuation of the Neolithic village in its recent isolation from the great urban civilizational centers in the marginal realms of the Earth such as Africa, the Pacific Islands, and large areas of North and South America until the arrival of the western colonial powers and their occupation of great areas of Paleolithic and Neolithic settlements after the fifteenth century C.E.

A third phase would be the continuation of the Neolithic peasant villages in their marginal status within the context of the greater urban civilizations as the source whence these urban centers obtain not only the resources in population and food supply but also inspiration for many of their artistic and cultural activities from the music, the oral literature, and the artistic qualities of these folk centers. These villages themselves achieve their creative energies from their abiding identity with the geographical regions and exuberant life expression of the natural world.

A fourth phase of the Neolithic village may well be its future revival as a model for the eco-villages that could become the new sustainable centers of the human venture as the infrastructures of the plundering industrial cities dissolve and human communities seek new models for a sustainable mode of human existence within the biosystems of the Earth.

I F WE LOOK back over the centuries since the rise of the Neolithic village we will see that the three earliest centers that were mentioned, Jericho, Hassuna, and Çatal Hüyük, each contributed to the primordial cultural and historical background for the future development of the western world and thereby established their relations with those larger destinies of the Earth itself that have emerged from this context.

Çatal Hüyük can be associated with the cultural area associated with the rise of Crete and its role throughout the Mediterranean, its influence on Mycenae, and its further influence in the Hellenic world. Jericho was later occupied by the Canaanites, who were a significant influence on the Hebrew peoples. Through them Jericho came eventually to be associated with the religious context of the western world. Mesopotamia, represented by Hassuna, Jarmo, and the other settlements in this area, was eventually associated with developments at Al Ubaid, Ur, Uruk, Eridu, Kish, and Nippur, in the lower regions of Mesopotamia where western intellectual development originated in many of its distinctive aspects.

Of these three it may be that the region of Mesopotamia needs to be noted especially in its influence over the western world and the destinies of the human. Located in the region known as Sumer in the lower marsh-land regions of the Euphrates River, the first efforts at irrigation took place in the early centuries of the fourth millennium B.C.E. This demanded social organization, an increase in population, and the capacity for designing a way of life integral with the natural phenomena of the region. In bringing this about Sumer took on a leadership role in the urban transformation of the human cultural process.

It was from Sumer and later Babylon in the lower regions of Mesopotamia that so much of our earliest western cosmological thinking originated, along with our mathematics and metallurgy. Our sixty-minute hour and our 360-degree circle originated here. Our earliest writing, our more conscious integration of the human process with the cosmological process, the early study of astronomy also took place here. The zodiac was identified along with its symbolizations. Legal codes and administrative processes later developed in the Roman period originated here. Different from Egypt and its centralized government by a Pharaoh considered as divine and possessed of total power, Sumer was a region of dispersed city-states with their own independent rule responsible at least in part to the people ruled. This political form would be later developed by the Athenians in the Greek world and passed on to the future. But if this is one decisive contribution of Sumer, the intellectual development is another of immense significance that has decisively altered the entire context of human-Earth relations.

FROM JERICHO TO Sumer, from 8000 to 3000 B.C.E., the great Neolithic revolution had taken place and these same cultural forces were pressing forward to their greater elaboration in the urban civilizations that were rising. That other regions too, Egypt and the Indus Valley civilizations particularly, were on the verge of this transformation seems to be clear from the future course of the basic Neolithic inventions. A threshold of change existed now with pressures building toward an horticultural-based city-centered mode of the human as successor to the Neolithic village.

At the transition from the Neolithic to the classical civilizations in the Near East there were apparently five to ten million humans on the planet. Within the next several thousand years the population would reach three hundred million people. As late as 1800 C.E. less than three percent of humans lived in an urban context. This itself indicates the power of the Neolithic peasant village to sustain itself even when a number of massive cosmopolitan cities had come into existence. In 1900 C.E. only ten percent of the world's population lived in an urban setting.

When we think of the classical civilizations of the past we might think of them as islands amid the sea of Neolithic settlements or as sparse centers amid a web of villages. Even into modern times India existed as a civilization of some six hundred thousand villages with few urban centers. China has remained a vast number of villages. The success of the Neolithic village might be estimated by its capacity to survive over the centuries, by its technological inventiveness in relation to its ecosystem, its artistic genius, its linguistic creativity, its oral literature, its song motifs and dance patterns, its intimate rapport with the spirit world, and by the brief space of time in which these villages gave birth to the urban civilizational forms and, until the rise of the modern industrial farm, has ever since sustained them in their food supply and even in their human population.

*Guardian Kings, Longmen Caves, Henan Province,
China, c. 675* B.C.E.

10.
Classical Civilizations

IN THE NEAR East and southeastern Europe, the Neolithic village reached a transition moment in the middle of the fourth millennium B.C.E. Population was increasing. A new economic situation based not only on agriculture but also on craft and engineering skills was beginning to function. Writing had been invented. Literature and the arts were expanding. More complex social structures were beginning to function. Changes were taking place in the religious orientation of the society.

The Neolithic perception of the universe in the vast course of its functioning, as originating in and cared for by the Great Mother deity, was changing. Throughout the Paleolithic and the Neolithic periods this maternal deity had presided in human consciousness as the ultimate term of reference in understanding the universe and in providing the psychic support for dealing with all those threatening forces in the surrounding environment. Since the development of agriculture the presence of the Great Mother as the all-pervasive creative principle of the universe had been enormously strengthened. So, too, in this period women participated extensively in the governance and in the general functioning of the society, as the source of its continued existence, the inventors of its basic arts, and also as artistic creators in the basic craft skills of the period.

For whatever causes, there was a change at the end of the fourth millennium B.C.E. from the dominant female deities to a dominance of male deities. The rise of technologies is also associated with the rise of men to dominant roles in the society.

In this context the Priest-Kings arose, rulers through whom the community attained its relation to the divine. Through the Priest-Kings came the divine directions appropriate for the society and a sense of the inner security it needed as it entered into the urban phase of its existence.

The Pharaoh in Egypt became the presence of the deity itself. In Sumer the ruler became the representative of the deity to the people and the representative of the people to the deity. In China the ruler received his mandate from heaven with the commission to bring about a comprehensive coordination of the society with the entire cosmological order. India was a society with extensive class differentiation, as well as ethnic and linguistic differences. The region never attained organized unity of political structure or expression until the time of Asoka in the third century B.C.E., who brought the various divisions of India together with Baluchistan and Afghanistan into a single political entity. As a world ruler or *cakravartin,* he fulfilled a cosmological as well as a political role. The Minoan civilization on the island of Crete remained oriented toward the Great Mother Deity throughout the period of its survival, until its destruction by natural disasters first in 1700 B.C.E. and again in 1450 B.C.E.

In every case the human social order was integrated with the cosmological order. These constituted a single order of the universe. Neither the cosmological nor the human was conceivable without the other. Each functioned in relation to the other while both functioned in relation to the presence of all-pervasive numinous power. Progressively the human ruler took on the attributes of the divine while the divine took on the attributes of the human ruler.

Meanwhile the primary technological inventions were pushing on to ever-greater power and productivity. The natural world as a resource for human use was being discovered. This period at the close of the fourth and throughout the first half of the third millennium B.C.E. was a decisive period in the advance of the human in its basic technologies. The technological achievements of this period remained unsurpassed in the various societies of the Eurasian world until modern times, and included wheel transportation, brick making, stone cutting, sailing vessels, sewage systems, water mains, mining and metallurgy, grain and food preservation, weaponry, carpentry, and architectural design.

Seed selection for improved grain production was taking place. Boats were invented early, along with the pole for use in shallow water, the oar and the sail for deeper waters. Animals were used for their wool and as food, as well as for their power in plowing fields. The ox-drawn plow culture region extended across Eurasia to China, Southeast Asia, and into Egypt, and was used nowhere else until modern times. Irrigation ditches were dug. Improved stone and later bronze blades were available for cutting trees and shaping wood. Chisels of stone or bronze and later steel were available for cutting and shaping stone.

While some bronze implements go back as far as 3700 B.C.E. the regular mining of copper began with the Egyptians in Sinai around 3000 B.C.E. Shortly afterward iron began to be mined, in 2800 B.C.E. Bronze was invented and used on an extensive scale, as was iron, in the form of steel for a variety of implements, especially for the short swords of the later period and for spearheads as well as for plows. Gold and silver were fashioned into sacred vessels and human ornaments. Clay was baked into bricks for building, and into ceramic ware intended for use as well as for decorative beauty. Travel routes were identified across the continents.

THE HUMAN WORLD and the wilderness were progressively differentiated from each other. Natural processes such as the random dispersion of seeds, the flowing of streams and rivers, and the freedom of animals were altered by human intrusion. Trade advanced within nations and between nations. Cargo vessels were on the seas and rivers and, later, on canals such as the Grand Canal from the Yangtze basin to the Yellow River to bring grain to the capital city. Caravans crossed from distant China to ancient Parthia where goods were taken on the Mediterranean and so to Rome as early as the first century C.E.

Meanwhile the landscape was altered by imposing architectural structures. In Egypt the pyramids were erected as eternal abodes for the pharaohs. In Sumer the terraced ziggurats were erected as sacred mountains where the deity and the ruler could meet in ritual rapport with each other. Along with such sacred places must be listed also the sacred palaces of the times. Of these the most magnificent was the Minoan palace of Crete built shortly after 2000 B.C.E. To the east in Persia we find the royal palaces of Persepolis with their lofty columns gleaming like a mirage in the desert. In Athens in the fifth century B.C.E., the Parthenon gave abiding expression to the Greek ideals of perfection. Rome found its ideal of political power expressed in the Forum and in the Coliseum, a calm expression of imperial power also expressed in the Pantheon.

The temples of India, built much later than the monumental structures of the Near East, were at times carved completely, inside and out, of massive natural rock formations. The Ming T'ang palace in China from the Shang period, 1550–1028 B.C.E., was erected as the central axis of the universe, where the ruler had only to sit on his throne and face south for the entire human society and the entire universe to revolve in their integral fashion, bringing the seasons each in their due moment. In Japan the temple-shrines of Shinto at Ise express the immediacy and utter simplicity of the numinous dimension of the natural world. The Japanese memorials to Buddha and the various bodhisattva manifestations portray the serene confidence of the Japanese in the powers that rule throughout the natural world. The Buddhist impact in Asia is further expressed in the immense cluster of ninth-century Buddhist temples at Borobudur in Java. Even more overwhelming are the twelfth-century shrines of Angkor Wat in Cambodia, basically Hindu in inspiration, but also including extensive sculptures relating to Buddhism, which cover some forty square miles and constitute a massive expression of religion in monumental architecture and exquisite sculpture in bas-relief.

The multitude of Muslim mosques in this period extends from Spain in the west to India and Indonesia in the east. In this medieval period Europe built its Gothic cathedrals with their towers rising above the horizon. In Meso-America the massive Olmec and Mayan and Aztec temples and pyramids identify the ultimate concerns of the indigenous peoples of this region, while the city and palaces of the Inca in Peru rest some seventeen thousand feet high in the Andes mountains as manifestations of political power on a cosmological scale.

The central cities of the various civilizations provided the power centers of human affairs in this transition period after 3000 B.C.E.: Memphis in Egypt, Babylon in Mesopotamia, Knossos on the island of Crete, Chandhu-Daro and Mohenjo-daro in the Indus Valley, Ch'ang-an in northern China, then the Rome of Augustus at the beginning of our era; then the Byzantine capital of the East Roman Empire of the twelfth century. In the Americas the sequence was Tikal of the Maya in Central America, Tenochtithlan of the Aztec, Cuzco, the Inca City of the Sun in the high Andes in the fourteenth century.

In such sites as these the divine and the human communicated with each other, ritually acted out the mysteries associated with the creation of the world, exchanged gifts, experienced conflict and reconciliation. These were the seats of government where justice was administered and the social order vindicated, the centers also that fostered the ideals expressed in the art, literature, and sacred books of the people. They educated the young concerning the world around them, the way of responding to the mysteries of the divine, the proper norms of human activity. These were the centers also of economic life, of trade and commerce for exchange of agricultural and craft products for the personal and institutional needs of the society.

Each of these civilizations has gone through cycles of development and decline, sometimes with a sequence of renewals, sometimes with successor populations, dynasties, and cultures. Each has left its impress on later generations. Innumerable monuments and whole cities have fallen into ruins and been forgotten even while the peoples themselves still carry in the unconscious depths of their minds these ancient traditions. In India the early civilizations of the Indus Valley civilizations disappeared and were forgotten from around 1800 B.C.E. Meanwhile the incoming civilization of the Sanskrit-speaking peoples and their civilization survived through the centuries, significantly influenced by both the Buddhist and the Islamic traditions that invaded the region.

China preserved its continuity from its early unification under the Shang dynasty through to the present with an irregular sequence of invading dynasties and return to the more traditional modes of expression. So also Japan, since becoming a literate civilization in the sixth century C.E., maintained its continuity as a center of creativity with an intense awareness of itself as a sacred center favored especially by Amaterasu, the feminine Sun-deity.

Beginning in the eleventh century B.C.E. the Olmec, Mayan, Toltec, Incan, and Aztec centers of civilization make their appearance in the Americas.

During these early centuries of the classical civilizations Israel was a small nation in Palestine with only a minor role in the political or cultural affairs that dominated the concerns of the massive empires in the surrounding regions of Eurasia and North Africa. Yet in the brooding course of history Israel would have an immense influence over the human story primarily because of its perception of the divine in historical events rather than in cosmological phenomena. Out of Israel would emerge a religious tradition that, allying itself with the thought traditions of Greece, the administrative genius of Rome, and the yet undisciplined energies of the barbarian invaders from the inner regions of the Eurasian world, would establish a complex of European power centers that would have enormous impact on the course of human affairs and planetary survival.

M IDWAY IN THIS five-thousand-year period of the classical civilizations, extending from 3500 B.C.E. to 1500 C.E., came the moment of deepest intellectual and spiritual reflection associated with, if not consequent on, a sense of the pathos of the human condition. The tragic dimension of existence established itself as a widespread and permanent concern amid all these accomplishments that we have presented. The transformation of human consciousness that occurred in the five hundred years between 800 B.C.E. and 300 B.C.E. brought about the spiritual visions that have determined the dominant sense of reality and value in the various civilizations ever since.

This sense of the tragic was something more than awareness of death, although this has been one of the more difficult experiences for humans to deal with. It was more an ontological or metaphysical perception of the transient nature of existence in the temporal order. It was perhaps a loss of innocence that existed in prior times when the human experience itself absorbed completely within the ever-renewing sequence of the natural world and unable to reflect on any other possibility open to the human.

This sense of the phenomenal world as oppressive to the more sublime aspects of the human, as a situation to be escaped from by spiritual discipline or some form of intellectual insight, constituted the ambivalence of the human situation. Life was full of entrancing sights and sounds and fragrance and feeling; yet it was also a threatening reality on every side. The visible world experienced as the shadow of some transcendent world was especially strong in Plato's writings. In India it was less the pain or the sorrow of the phenomenal world than a sense of its unreality that required a liberation experience into the world of the absolute.

THIS SENSE OF sorrow found one of its earliest expressions in the story of Gilgamesh, the heroic figure in an early Babylonian epic who is so shaken by the death of his friend Enkidu that he develops an acute fear of death and searches for some way to immortality. After obtaining the plant that bestows eternal life, Gilgamesh loses the plant and is told that only a gloomy future awaits humans after death. Earlier, in Egypt, in the period after the end of the Old Kingdom around 2400 B.C.E. an extensive pessimism developed among the people. The centralized source of order had lost its efficacy. Social turbulence was being experienced for the first time on this extensive scale. The entire functioning of the universe was seen as a declining situation.

The need for liberation, not simply from death but from temporality itself, found expression in Moksha in the Hindu world. In Buddhism liberation from the bond of karma and from the deepest sorrow of life, Dukkha, was found in the nirvana experience. In China the experience of human failure led to unending discussion of the evil in the world of the human whether or not humans were inherently wicked. In a poignant poem written in the fourth century B.C.E. a court minister, Ch'u Yuan, gives expression to the deepest sorrow of existence in a poem entitled Li Sao, "Falling into Trouble." Mencius tells of the tendency in humans as they move out of childhood to throw away their minds. Thereafter the whole of life is to recover the lost mind of the child. The way of proper conduct in China is to be sensitized to the first subtle spontaneities within the heavenly bestowed nature of the human, for here resides an infallible guide.

In the Greek world this issue of the human condition is presented in all its full force in the ritual Tragedies of Aeschylus, Sophocles, and Euripides. The death of Socrates struck such an impression in the mind of Plato that thereafter he spoke of the philosopher as one who is always dying to the visible world to gain an immediacy with the transcendent world. For Plato the absolute world is found in the Agathon, the Good. There in the Agathon are the true realities, the transcendent forms of all things, forms that are reflected here in the shadow world of time and transience. The direct vision of these forms in their numinous reality is available for humans who are able to move from the beauty of the sense world to the beauty of the inner psychic world to that which is beauty in itself.

Later Plotinus the Neoplatonist established his teaching of the Supreme One whence all existence flows. Beatitude is presented as unification with this divine reality beyond any mode of the existing world, beyond imagination, beyond thought itself; beyond all this is the ecstatic experience whereby the human transcends even itself in a state of abiding bliss.

In the world of Israel concern over the human condition found its most elaborate expression in the Book of Job where the sufferings of the good person are seen as a basic anomaly from which the only conclusion is that suffering cannot be equated with any explanation available to humans. But if there is no rational explanation there is the overwhelming grandeur of life. This should provide sufficient evidence that benign forces are ultimately at work throughout the universe.

Christianity dealt with this tragic dimension of life through the doctrine of some primordial failure of the human that resulted in a profound alienation from the divine. Recovery from this alienation required a corresponding redemption experience accomplished by a divine redeeming personality in historical rather than mythic cosmological time.

These spiritual experiences at the center of the traditional civilizations were communicated through sacred revelations that took place in the depths of the human unconscious and were awakened by inner dreams or outward experiences, and found abiding form in sacred scriptures. These revelations appear in the Rishis, or Seers, of ancient India, in the experience of Buddha in Bodh Gaya, in the illuminations of the Persian Zoroaster, in the biblical prophets, in the Christ appearance and the Pauline Epistles. Here the sense of reality, the norms of good and evil, were established in a variety of enduring expressions available for everyone.

THE SUPREME INVENTION of this period, with immense influence on the course of human history, was the invention of writing, first the Sumerian cuneiform and later the Egyptian hieroglyphics and eventually the alphabet. By this means thought, ideas, dreams, and revelations were fixed in an abiding fashion. Even without an alphabet, China and the entire East Asian world were able to record their thoughts in a writing system based on formalized pictographic and ideographic characters. The sacred scriptures of each of the traditions became normative for their sense of reality.

These visionary experiences that took place in mythic archetypal modes were articulated in mythic and historical narratives, in ritual, and in wisdom reflections. Then came the more explanatory works, the commentaries on the sacred revelations, and the mystical disciplines.

Afterward came the time of synthesis. This constitutes the "medieval" period of the various traditions. This occurs in the Hindu work entitled the *Brahma Sutra* and its commentaries over a thousand-year period, the first all-inclusive synthesis of Hindu thought. In Buddhism such a synthesis is found in the *Abhidamma,* the *Path of Purification* of Buddhaghosha, the *Verses of Nagarjuna,* the *Lotus Sutra,* and the *Vimalakirti Sutra.* In the Chinese world the Neo-Confucian synthesis was articulated most fully by Chu Hsi. In the Christian world, after the Fathers of the Church come the Scholastics, Scotus Erigina, Peter Lombard, Thomas of Aquin, and Duns Scotus writing their systematic treatises on the whole of the belief structures of the Christian world. The great synthesis of Islamic thought was achieved by al-Ghazali in the eleventh century C.E. Moses Maimonides in the twelfth century C.E. was the principal exponent of Jewish thought.

Also in this period the various centers began the universalization process that has continued through the centuries. There was less and less isolation. Mutual influence began between Buddhism and Chinese thought. Buddhism interacted so extensively with earlier traditions throughout the entire East Asian, Southeast, and South Asian worlds that it might be considered the only pan-Asian unifying principle justifying the use of a single term for this extensive area of the continent. The personality of Buddha, his image, and the pagoda shrines of Buddhism are found everywhere throughout this area.

Biblical religions interacted with the Greek world, its mystery religions, its deities, its philosophical thinking. Then Christianity passed on to the tribal peoples who entered Europe in these later centuries of the Roman Empire and throughout the period from the sixth through the eleventh centuries. Later, in the seventh century C.E., Islam began its movement both east into what is now Pakistan, China, Sri Lanka, the Malay peninsula, and

Indonesia, and west across North Africa into Spain, thereby establishing itself in a southern band across the entire Eurasian continent, with an extension across the top of the African continent. In its movement eastward, in the Philippines, Islam met the westward course of Christianity carried by Magellan into the Pacific regions in the sixteenth century C.E.

The early civilizations of the Western Hemisphere, the Olmec, Mayan, Toltec, Aztec, Inca, Pueblo, these were beyond the range of Eurasian communication throughout this entire period. Thus there is no clearly established exchange of goods or ideas or spiritual influence between any of the pre-Columbian American and any of the Eurasian cultures. The break in continuity with the Eurasian world had taken place long before, since there is still no adequate indication of linguistic relationship. The geographical and climatic conditions that had originally enabled these indigenous peoples to come here no longer allowed them to keep in communication with their distant origins.

THE OVERALL RHYTHM of development in the immense Eurasian continent can be identified in the pressures from the warring pastoral peoples of the steppe regions of inner Eurasia upon the agricultural societies in east, south, southwest, and west Eurasia. This outward movement occurred in three distinct periods. On each of these occasions significant changes were wrought throughout this vast region.

A first outward movement in the middle of the second millenium B.C.E. flowed over these outer regions and then subsided, but not without leaving immense changes with its outward movement in the various directions. This first movement determined to an extensive degree the linguistic regions of the vast continent. Specifically, it brought about the spread of Indo-European languages from the Celtic borders in the west to the Baltic states in the north, to the Slavic realms of Eastern Europe, on to the Greek-Latin world of the Mediterranean, and on to the east with the Sanskrit language of India.

A second movement occurred later, at the height of the Roman Empire in the early centuries of our era. It reached its culmination in the reign of Diocletian at the end of the third century C.E. and caused the succeeding emperor, Constantine, to move the capital of the empire to the city of Byzantium, which he renamed Constantinople. The forces of the inner Eurasian world moved outward once again to invade the entire range of countries from China in the east to India in the south, to the Mediterranean regions of southwest Eurasia, to Europe in the west. This tide of people broke through the *limes,* the famous wall of Rome extending from the Danube to the Rhine, which was massively fortified at this time.

Rome fell to the Goths in the year 410 C.E., an event that shook the European and Mediterranean worlds as though the end of time had come. The Greco-Roman heritage had so possessed the souls of people throughout the entire region that life beyond this established context was inconceivable. The most renowned effort to assimilate the meaning of this event was that of Saint Augustine, who wrote *The City of God* between 410 and 424 C.E., a work that presented the story of the world within the biblical context and enveloped the fall of Rome in the larger story of the universe itself. This was the book that, more than any other, carried Christians through the civilizational decline of the Dark Ages into the new creative period, beginning in the twelfth century, when the cities of Europe began to rise once again as centers of life, thought, and creativity.

For a third time, in the thirteenth century, the tide of forces within Eurasia gathered under the leadership of Genghis Khan, the Mongol warrior. These tribes entered history with devastating military power and yet also with political genius, moving west across three thousand miles of the Eurasian continent into the regions of Poland and Hungary; also moving east to conquer and rule China for more than a hundred years, from 1259–1368 C.E. For over two hundred years the political forces introduced by the Mongols determined a large part of the political structure of the Central and East Asian regions. These Mongol forces even sought to attack Japan across the sea from China. They were stopped only by the kamikaze, the divine wind that sank the Mongol fleet.

E ARLIER, IN THE ninth century, Christian Europe, with its center as somewhat of an island in the south of France, had been assaulted by the Normans coming down from the north, the Magyars from the east, and the Moslems from the south. In consequence of this threatening situation the Crusades were declared in 1095 and the world movement of the European peoples was begun. After two centuries of crusading, Europe went through a sequence of dynastic wars in the medieval and renaissance periods. Then in the fifteenth century, feeling the Islamic threat after the Turkish conquest of Constantinople in 1453, Europe sought to diminish the threat by a simultaneous Christian assault on the Muslim forces from both the east and the west, since it was known that there were Christians on the other side of the Islamic world.

This assault was to be carried out by establishing contact with these distant Christians and then attacking the Muslim forces from both these positions. Stirred largely by this motive Europeans sailed down the African coast around the Cape of Good Hope and on into their first contact with Ethiopia, India, and from there on to the East Asian world. They never realized their purposes, however, because of their inability to communicate effectively with the Coptic Christians in Ethiopia and the Malabar Christians in India, but also because the commercial preoccupations of the Europeans took precedence over their military concerns.

Just prior to the expeditions that reached India, in the closing years of the fifteenth century, Columbus ventured across the Atlantic and established the first enduring European contact with the American world. Henceforth everything was changed. These discoveries and conquests, along with new scientific and technological discoveries, enabled a new cultural order to emerge which these earlier civilizations could not have managed from within their own resources.

This inherent drive of humans to establish contact with each other for trade relations and political alliances, as well as for religious and spiritual purposes, is one of the most consistent aspects of human activities throughout the course of these centuries. Already in the Han period, in 138 B.C.E., Chang Ch'ien traveled west from China to discover a southern route to Bactria. Later in this same period Chinese expeditions went west all the way through the Sinkiang region to Parthia to establish relations with that state.

As early as the first century C.E., Indian Buddhists made the long journey through the mountains of Central Asia to Lo-yang in China for both trade and religious purposes. In response a long list of over two hundred Chinese Buddhist pilgrims traveled at different times to visit the holy places of India, culminating in the remarkable journey of Hsuan-tsang in the year from 629 to 645 C.E. to India and to the greatest learning center in Asia of this period, the Buddhist University at Nalanda.

From the west the most remarkable traveler was Marco Polo, who journeyed from Acre on the eastern shores of the Mediterranean across to the Oxus river, then on through the Pamirs to Kashgar, Yarkand, and Khotan to Lobnor at the eastern end of the Tarim basin in western China. From there he ventured on across the Gobi desert to Karakorum where finally he met with Kublai Khan. After some twenty years in the service of the Khan in China, Marco journeyed by sea to Sumatra and India, then on to Persia, and finally returned to his homeland, Venice, in 1295.

In the period 1325–1355 C.E. the greatest of pre-modern explorers, the Muslim Ibn Battuta, journeyed throughout the Middle East, Arabia, Syria, and Persia, on through India to Sri Lanka, Sumatra, the Maldive Islands; then afterward to the west across North Africa, then south to Timbuktu and the Niger.

Throughout these centuries efforts were being made to establish some functional rapport between the various political and cultural traditions. Genghis Khan and his successors, however terrifying their assault on the peoples of the entire Asian and east European worlds, ruled for a while over the largest and most diverse grouping of peoples thus far in the course of human history. Genghis Khan himself was aware of the religious and cultural diversity of the peoples of this vast region of central and eastern Eurasia and sought to establish a functional rapport among their various cultural and religious traditions.

Nothing seems to intensify these inner tensions and the resulting warring process among peoples so effectively as their various religious and spiritual commitments. In the western world and in all the biblically derived monotheistic religions, this difficulty in relating to other traditions has been especially strong. Since religion was such a powerful influence over all phases of individual and social life this was a major force in producing the human situation at the end of the fifteenth century C.E.

Whatever the causes, the warring process went on endlessly. The deities themselves were warring figures. This can be vividly seen in Indra, the principal deity of the Vedic hymns, a warrior deity in whose honor more hymns are found in the Rigveda than for any other of the deities. Yahweh of the Bible is a local warrior deity elevated to the status of transcendent monotheistic deity of the biblical peoples. Zeus of the Greeks, Thor of the Germanic tribes: all these were gods of war. In China the endless internal conflicts and engagements with the assaulting forces of the surrounding tribal peoples all give evidence of the warring situations that existed throughout this period. In Japan the warrior ideal has captivated the imagination ever since.

The wars fought by humans were frequently wars between factions within the societies themselves. The Chinese warring process in the centuries between 700 and 221 B.C.E. was a war to decide which dynasty should rule China. This was one of the bloodiest periods of any people and any period. It involved the death of hundreds of thousands of people. Later came the wars of the Chinese with the outer tribal forces that were both dependent on and in tension with the empire.

One of the unique causes for war was the commitment of the Aztec to warring for the purpose of obtaining victims for the human sacrifices offered to the Sun-deity, Huitzilopochtli. Only in this manner was there any assurance that the sun would remain shining.

In India the epic story of the Mahabharata tells of the long series of battles for the rule of the society.

In the middle of the fourth century B.C.E. Alexander established control over the eastern Mediterranean and moved farther east into the valley of the Indus river. His empire was succeeded by the Romans who first, under Caesar, brought southern Europe and Britain into the empire. Shortly afterward the empire was extended throughout the Middle East and then into Egypt. Then came the tribal conquest of much of the imperial territory in the first five centuries of the present era.

So too in China various tribal forces to the west, northwest, and northeast first brought the Han dynasty to an end in the third century C.E., then disrupted the T'ang rule of China in the mid-eighth century, then took over the rule of China from the Sung Dynasty in the thirteenth century, and terminated the Ming dynasty in 1644 C.E.

BEYOND THE ISSUE of cultural relatedness there is the issue of the relation of these civilizations with the biosystems, the geographical formations, and the seasonal cycles whereby the planet functions in some coherent way.

During the five thousand years under consideration, the relation of the human species with the other species of the planet varied considerably in the various civilizations. In the Hindu world there has been a deep feeling for the identities between the human and the natural world. This can be seen especially in the pervasive sense of compassion for every living being. The Jains take this to such extremes that they take great care not to harm even the lowest forms of life.

If the other traditions are less sensitive on this point, they do emphasize that the compassion due to humans should be extended to other forms of life. This attitude is quite strong in Buddhist tradition. In one of the Jataka narratives Buddha throws himself over a cliff to his death in order to feed a tigress with her starving cubs. Shanti Deva, one of the more illustrious Buddhist writers of the fifth century C.E., offered to take upon himself all sufferings of all beings in order that they would attain the blessed release from the karma impeding their liberation into nirvana, in the conviction that it was better that he alone should suffer rather than that the great multitude of living beings should continue to exist in misery.

In China the Confucian teaching has generally been toward a benign concern for the nonhuman world, a teaching found in Mencius with his concern that the birds not be disturbed during their nesting season and that the nets used in fishing not have too fine a mesh. In one of the most important chapters of the *Book of Ritual*, entitled in English *The Doctrine of the Mean*, the teaching is communicated that the human through complete authenticity establishes itself as a Third with Heaven and Earth in bringing all things into being and carrying them on to their completion. This teaching is presented also by Mencius, considering that "all things are complete within us." This mystique is found throughout the course of Chinese thought, in its poetry and especially in its landscape paintings.

After the arrival of Buddhism this mystique was intensified in China, as we can see in the sensitivity of the twelfth-century administrator, Chou Tun-i, toward the natural world; a sensitivity that would not permit him even to cut the grass surrounding his dwelling place. A more complete expression is found in Wang Yang-ming in the late fifteenth century where he speaks about human consciousness of Heaven and Earth and the Myriad Things as "One Body." Evidence of our identity with the One Body of the universe he indicated through the pain that we experience when we witness pain inflicted on any other being. Only those who have this experience of the identity of the human with the entire order of things can be considered as possessing a completely human mode of being.

But while there was an extensive teaching concerning the intimate rapport between the human and the natural world, in the practices of the various peoples there was extensive devastation throughout the various civilized regions of the Eurasian world, and to some extent in the Americas. China is a foremost example of theoretic intimacy with the Earth by a people who seemed not to understand the harm they were doing in denuding the country of its forests to obtain larger areas for cultivation. In losing their forests they lost much of the fertility of their soils, although due to their understanding of returning waste to the soil they did manage to keep enough for survival. Over the centuries, however, immense quantities of fertile soils have been washed out to sea.

Similar statements concerning the erosion of soils and consequent desertification can be made concerning other regions of the world, especially North Africa, once an extremely fertile region that has now become desert. In Greece, Plato mentions in the *Critias* that the hills had been cleared of their trees and the springs consequently had dried up. To what extent these destructive activities came from ignorance and to what extent from insensitivity to natural life communities can be questioned, but the consequence remains through the centuries.

In the western spiritual traditions nature was seen originally as created by the divine and as a primary revelation of the divine, even as a revelatory experience intimately related to the biblical revelation. The psalms included the various natural phenomena within the sacred community. Hildegard of Bingen, in the twelfth century, perceived the divine primarily in the greening of the Earth. Also much of western identity with the natural world was from the stoic tradition, which envisaged the universe as a great Cosmopolis in which every being held citizenship. At a deeper mystical level the neoplatonism of the West provided the basis for a comprehensive community of the universe in which, as Dionysius says in *The Divine Names,* "All things have a friendship relation with each other."

Yet within all these Eurasian traditions that exalt the integral nature of the universe the course of human-Earth relations has never been without its tensions. As the technologies of the human have advanced so have the demands made on the Earth. This stress began in the Paleolithic period when humans assisted in bringing about the extinction of many of the larger animal species. This may have been due primarily to other causes such as climate changes, but the human assault on these species was a severe one and made the extinction more inevitable.

Not only in the West but on broad human scale there developed an exaggerated anthropocentrism. In the West this was based to some extent on an inadequate understanding of the Greek saying that the human is the measure of all things. In the West also there was the ever-increasing awareness of the pathos of the human, the need to fulfill the obligations associated with divine-human and inter-human relations. But this so exhausted the energies and attention of the society that the natural world gradually faded from consciousness as a significant concern.

This anthropocentrism was altered to a theocentrism when the Plague struck Europe in 1347–1349. A third of Europe died. Since at this time there was no explanation of this event in terms of germ theory, the general conclusion was that there was too great attachment to the Earth. The great need was for spiritual detachment and absorption into the divine. This led to a more absolute commitment to salvation from the Earth rather than to an integral relation with the Earth as a single sacred community.

In the West especially, the mystical bonding of the human with the natural world was progressively weakened. Humans, in differing degrees, lost their capacity to hear the voices of the natural world. They no longer heard the voices of the mountains or the valleys, the rivers or the sea, the sun, moon, or stars; they no longer had a sense of the experience communicated by the various animals, an experience that was emotional and esthetic, but even more than that. These languages of the dawn and sunset are transformations of the soul at its deepest level.

By the year 1500 C.E. western civilization had already lost much of this earlier experience and was beginning to influence the larger world in the direction of an anthropocentrism or a theocentrism that negates the intimate unity between the natural, the human, and the divine worlds. Earth itself was no longer seen as a communion of subjects. It had become a collection of objects to be adjusted to in an external manner.

ASSOCIATED WITH THIS insensitivity to the natural world was an insensitivity throughout this civilizational period to the women of the community. A patriarchal dominance was everywhere evident. Both the political and religious institutions of this period gave expression to patriarchal ideals oppressive to the integral development of women. The exclusion of women from the public life and institutions of the various societies came to be an integral part of the religious teachings and rituals of the period. Just how profoundly these effects were felt we are only beginning to understand. The traditional civilizations were shaped in the context of the unceasing toil of women, their servant status, their possession by men, and their role as pleasure-givers for others. In all this they were profoundly depersonalized.

Another difficulty throughout this period was the subordination of some groups by others within the same social unity. Slavery had existed on an extensive scale from Neolithic times. Frequently those taken as prisoners in warfare were reduced to slave status. This we witness especially in the classical Mediterranean world, in Greece and Rome. So too in earlier societies there were submerged classes that functioned as slaves even when this could not be considered their proper designation. This can be observed in the Sudra class in India, an outcast group. The various castes of traditional India were originally constituents of an integral social order with complementary services rendered to each other. But then this evolved into a rigid subordination of some groups by others. That lower social status was considered the result of moral delinquency only added to the difficulty.

Even besides the social status, the assumption of power by elite groups was practically universal. The privileged classes, those who enjoyed the higher status, were the professional religious persons, the ruling authorities, the wealthy traders, and the landowners.

Another aspect of this period is the rise of the human population throughout the world. In the early Neolithic period, around 9000 B.C.E. the human population of the world was apparently a million or more persons. By the end of the third millennium B.C.E. the human population was possibly five to ten million. Then by the first century C.E. the population of the world is estimated at three hundred million. In the late medieval period there was a significant decline throughout the entire Eurasian continent due to the plague that apparently came from the Asian world to Europe. The population of China dropped in the early thirteenth century, it seems, from around one hundred and twenty million to sixty-five million. In Europe perhaps a third of the people died from the plague that had its greatest impact between 1347 and 1349 C.E. Yet there was a recovery within a century and a half so that by the year 1500 C.E. the population was somewhere between four and five hundred million.

This represents a sizable population for the planet, yet something that the planet was able to sustain with a certain ease. Even so, the local centers constantly experienced populations beyond what could be managed within the available resources. The main limitations on population were due to food supply and disease, although humanly devised methods of birth limitation existed from an early period.

The massive centers of population were in China, India, and the Near East up through most of this period. The European population up through the medieval period was not comparable to the extensive population of China at this time.

THE VARIOUS CIVILIZATIONS dealt with thus far had taken extensive but limited control over the planet by 1500 C.E. The larger part of the Earth remained in its Paleolithic or Neolithic status as far as its human population was concerned. The period of civilizational transformation of the planet was also the period of development for the indigenous peoples of the world. Although these peoples never shared in the more complex modes of civilizational expression, they kept and further developed their way of life deep in communion with the larger life systems of the Earth, with the spirit world of the mountains and rivers and valleys, and with all those living forms that exist on the planet.

In the year 1500 C.E. the larger part of the Earth—sub-Saharan Africa, North America, South America, Australia, the entire South Pacific, the inner regions of the Eurasian continent—remained outside the influence of the civilizations we have discussed so far. These regions included a multitude of indigenous peoples with their thousands of languages and immense diversity of customs, their arts and poetry, their songs and dance and religious rituals, their laws and governance. Such diversity in the display of human magnificence the world would never see again, for this was a transitional period; influences would flow over the planet that had never been dreamed of previously. Even when the tide of European peoples moved over the Earth after 1500 these tribal peoples remained in great numbers throughout every continent except Western Europe.

The area of indigenous peoples, up through the period we are presenting, included most of Africa below the Sahara, a region covered with luxuriant vegetative growth, amazing diversity and abundance in its animal life, with its great herds of elephants and feline creatures and birds, and its various species of fish. In Africa at this time there were an estimated three thousand distinctive human groups, with perhaps one thousand languages. Most of these peoples were small farmers living a Neolithic village existence.

Some portions of North Africa just below the Sahara had already experienced the influences of Islam, and the first kingdoms had been established. Ghana, in pre-Islamic times, in the fifth and sixth centuries, was the earliest kingdom. In the thirteenth century Mali, an Islamic-influenced country, was taking on more developed forms of social and cultural development.

The indigenous cultures developed throughout Africa manifest a unique exuberance in their music and drawings, and sculptures, and dancing. There is such extroversion in the tribal forms of drumming, dancing, and in all forms of artistic expression in this region, all so different from the introversion of the yogic traditions of India. Spiritually, the African peoples have an awareness of a High God Creator of the world. But the peoples are generally more in contact with the spirit world that is found in the elements, the air, the water, in all natural phenomena. Each of the various peoples has its own cosmology expressed in their narratives that tell of the origins of the world, of all its inhabitants, especially the human.

The other vast region of the planet not even known to the peoples of the Eurasian civilizations in the period under discussion was the Americas. This region, designated as the Western Hemisphere, contained an enormous amount of the vegetation that enabled the planet to function in the larger dimensions of its biosystems. A fantastic number of animals roamed the forests, the jungles, the mountains, the plains, and the waterways.

Here, in both the northern and southern continents, there were perhaps fifty million indigenous peoples.

The civilizations of the Maya, the Aztec, and the Inca appeared in the Americas and reached their high moment of development prior to the European invasion of these continents. But still the vast reaches of the Americas were touched by these centers in only a limited manner, except for the plants that were domesticated and cultivated for food in these regions. Among the high cultural achievements of this period was their highly developed agriculture. Probably the earliest agricultural inventions in this hemisphere took place around 6000 B.C.E.

WHEN WE OBSERVE the entire complex of civilizations that emerged in this five-thousand-year period we are struck both by the common and by the distinctive features of the various traditions. What they have in common is an agricultural basis, religious rituals, governments integrally related by ceremonial observances to the cosmological sequence of time, monumental architecture, the arts of weaving and pottery-making, generally some system of recording events, and even literary creativity.

These cultures are best identified not as higher or lower but as differentiated modes of cultural expression according to their integration with the functioning of the natural world, their capacity for specialization of function, the effectiveness of their social organization, their capacity for urban existence, their educational adequacy for the established form of life, and their capacity for ordered systems of knowledge.

In each of these civilizations we can observe a distinctive form that gives to each its identity. In the Chinese world we find that the basic orientation is toward a harmony with the deepest rhythms of the universe. This is what led the Taoists in China to their emphasis on the primary causality of the Tao rather than the secondary causality of the human agent in accomplishing any activity. The entire educational program was to cultivate a sensitivity to the inner spontaneities of the Tao as manifested in the individual persons. In this manner the person attains an identity with every being in the universe. Only when we experience ourselves as "One Body with Heaven, Earth, and the Myriad Things" can we consider ourselves as possessing a fully achieved humanity. Here the cosmological order is the primary referent.

This is vastly different from the Hindu world where the basic spiritual emphasis of the culture is an awakening to the transcendent realm of that Supreme Reality which is beyond all knowing and where there is a feeling for the unreality of the entire visible world, the world of Maya. Then too there is the feeling of the world of the Absolute, the world beyond all multiplicity. Here there is an emphasis on *Neti, Neti;* not this, not that. This experience receives further emphasis in the Buddhist experience of nirvana. This experience is much more metaphysical than cosmological.

In China precise chronological records were kept from a very early time, and we know precisely when historical events took place from the eighth century B.C.E. onward. In India, on the other hand, we cannot date anything with great precision out of India's own records until recent centuries. Even then our dating is generally dependent on outside contacts with India. So we are dependent on the dates of Alexander's invasion of India in 327 B.C.E., for a precise date of this period. Then we can date from the sequence of Chinese Buddhist pilgrims who went to India from the third until the ninth century C.E.

In the Greek world we have a consistent emphasis on the reality of the human psyche and its capacity to interact with the external world. We also have the development of a dialectical reasoning process.

One other tradition that needs to be indicated here is the western biblical Christian tradition in its relation with the classical humanist traditions of the Greek and Roman worlds, later with the tribal peoples that invaded Europe during the first ten centuries of our era, and also with the Muslim traditions brought to the Mediterranean and Spanish worlds from 600–1500 C.E. If the integration of this diversity of cultural forces was an immense challenge, it was correspondingly an exceptional source of intellectual insight and cultural energy.

These traditions brought with them not only their explicit teachings but also an array of esoteric teachings allied with the hermetic books, alchemical teachings, Neoplatonism, Neopythagoreanism, and Neostoicism. Only in the fifteenth century, after the occupation of Constantinople by the Turks in 1453, did the western peoples have the complete corpus of the Plato's dialogues, brought by Greeks who fled to the west. All of these sources carried a wealth of new images and symbolic interpretations of the universe that greatly enriched the imagination and stirred the intellect of renaissance scholars. This was a turbulent period in western history that broke apart so many of the fixations that had taken place in western thought and imagination. It made possible also the transition to our scientific method of inquiry into the structure and functioning of the universe about us.

Even more powerful in the world-shaking dynamism of this period was the historical realism of the west, allied with its commitment to the rational processes of the human mind and to the inherent value of the human person. This had run its course to some extent in its medieval expression that was passing away at this time. But it was ready to enter onto a new phase of its destiny.

Seaman's Atlas, Amsterdam, 1684

11.
Rise of Nations

I N THE SIXTEENTH century all the major civilizations of the Earth were
at a high moment in their political and cultural achievement—however
oppressive in terms of their patriarchal mode and their exploitation of the
multitude of peoples who bore the burdens and endured the cruelties of
their wars, their social indignities, and their economic establishments.
China was in the grandeur of its Ming period with its ceramic art and novel
literature and Neoconfucian intellectual life. India was experiencing the
florescence of the Moghul dynasty, which reached its height in the reign of
Akbar (1556–1605). This was the century of Suleiman the Magnificent
who ruled the Ottoman Empire (1520–1556) from Constantinople. Russia
after 1480 was recovering from two and a half centuries of Tatar occupation and entering on its imperial phase. In western Europe a new vigor was
being asserted by the monarchies in Spain, France, and England. Spain
had just expelled the Muslim armies and was able to establish its national
unity by joining the two regions of the country under Ferdinand and Isabella. England was entering its seagoing period that would continue its
expansion until its empire had encompassed the Earth at the end of the
nineteenth century. In the Americas the Inca and Aztec empires were at
their most brilliant moment just before they were seized by the Spanish
conquistadores.

At this moment Europe was recovering from the experience of having
itself been endangered throughout the centuries since the decline of the
Roman Empire. The ninth-century invasions of the Normans, the Magyars,
and the Muslims were over. The crusades had come to an end. Yet after the
fourteenth century, when these assaults were terminated, the European

world felt threatened again when the Turks took Constantinople in 1453. Prior to this time Henry the Navigator of Portugal had begun sending ships on explorations down the African coast with the intention of outflanking the Muslim forces by establishing contact with Christian forces in Ethiopia and Asia and coordinating a simultaneous attack against the Muslims from both east and west. This goal was never achieved due to prior economic concerns and lack of spiritual rapport with eastern Christians once contact was established.

Vasco da Gama reached the Malabar coast of India in 1498. Before the end of the century Columbus had reached the continents of the Western Hemisphere, in 1493. During this same fifteenth century the Chinese, with superb seagoing vessels, were sailing through the Straits of Malacca past the Indian subcontinent to the Persian Gulf on trading and tribute missions under Cheng Ho. The last of his seven voyages to the Philippines, Java, Sumatra, Ceylon, and the Persian Gulf was completed in 1433. Afterward China decided to withdraw from its sailing ventures and turn landward. In 1581 Russia crossed the Ural mountains and by 1640 had reached the Pacific coast, a period of less than sixty years. At this same time the Spanish, French, Dutch, and English had reached the eastern coast of North America and established their first settlements: Spain founded St. Augustine in 1565; the French founded Quebec settlement in 1608; the English founded Jamestown in 1607 and began their journey to the Pacific, a journey complementary to that of Russia, which moved much more rapidly to reach the Pacific, crossing over to Alaska by 1784 and then on down the coast to establish Fort Ross in California by 1812.

India had already established its presence through traders and religious personalities throughout southeast Asia and along the east coast of Africa. The Chinese had extended their empire into the south and western parts of the Asian continent. Japan was just entering its century of civil strife to determine which of the regional lords would unify the country, a strife that ended with the establishment of the Tokugawa dynasty in 1600 by Ieyasu. Under his successor, Iemitsu, in 1623 the policy of national isolation was adopted to last until the mid-nineteenth century.

Following Prince Henry the Navigator of Portugal, Ferdinand and Isabella of Spain supported the overseas ventures that resulted in extensive Spanish occupation of Central and South America. Following Portugal and

Spain, the coastal European powers turned seaward, and this has made all the difference. European peoples through colonization soon occupied a good portion of the planet: North and South America, Australia, and New Zealand. When the changes were finally over, the political-cultural shape of the planet had been altered in a radical manner. Humans had circumnavigated the planet. The various human communities were in contact with each other and turned toward a common destiny in a manner never previously existing. Yet this new wandering of European peoples could hardly match the wanderings of the earliest peoples when the planet was first occupied, or even the wanderings that brought the Cro-Magnon peoples into Europe some forty thousand years ago, those that brought the tribal peoples into Australia, and those that brought the American Indians into the American continents. So these centuries of European occupation of the planet can be considered simply as a continuation of an age-old human process.

ONE OF THE MOST significant ventures of this period was the British military-political-economic occupation of India and the gradual takeover of this country from its other European occupants, the Portuguese and the French. This was done through the British East India Company, an instrument commissioned by the British government and supported by military troops for accomplishing its purposes. These purposes were control of territory and monopoly on trade, both of which were tragic for India, ruining its weaving and craft industries, although of course they benefited both the British empire and the commercial profit of the Company. Although this arrangement functioned adequately for a while, there was a final battle fought by the British under Robert Clive against the Nawab of Bengal. The treaty arrived at after these conflicts in 1757, the Treaty of Plassey, determined the extended control of the British over India and was officially recognized by the other European countries in the Treaty of Paris in 1763 when Spain, France, and Britain agreed on the distribution of the colonial world of that time among themselves.

European peoples and their colonial realms had been in contact with Africa through the slave trade since the seventeenth century. Yet not much was known of the interior of Africa until the journey of Livingstone across

209

the continent from Zambezi to Luanda. Richard Burton, the British explorer and unofficial government agent, was also in Africa from 1858 to 1859 when he discovered Lake Tanganyika. The continent was extensively explored by explorers from several nationalities. Among them were the Englishmen John Speke, and James Grant, and the Germans Georg Schweinfurth and Herman Von Wissmann. Others, from Italy, included Pellegrino Mattecucci and Alfonso Massari.

Until the nineteenth century the European countries interested in Africa were settled in trading posts along the coasts. These were mainly Spanish, Portuguese, Dutch, and French. Britain held only a few isolated places in the sub-Saharan region. Until 1883 Germany had not entered the colonial venture in Africa. The Muslims were present throughout north Africa and along the eastern coast of the continent.

In 1884–1885 the Berlin Conference was convened with all the European powers and Japan to consider the western entry into West Africa. Fourteen nations, including the United States, attended. The Congo Basin was to remain open for trade and travel.

By 1914 Britain, France, Italy, Portugal, Germany, Belgium, and Spain had divided up Africa with minimal consideration for ethnic, linguistic, or natural boundaries. The French were mainly in West Africa, the British in Egypt, the Sudan, and along the eastern coastal region as well as in Rhodesia and South Africa. The Germans were in southwest and southeast Africa, Belgium in the south-central region.

In the late nineteenth century the only extensive region that remained outside the control of European powers to any extent was East Asia: China, Korea, and Japan. Mongolia was being fought over by the Chinese and the Russians. Siberia had been occupied by Russia in the seventeenth century. Japan was forced by the Perry mission from the United States to open its country to western trade in 1854. China, after being forced by the British in the Opium War of 1841 to establish trading ports with the western nations, experienced fifteen years of conflict between 1850 and 1865 when the Taiping rebellion was carried out against the Manchu dynasty to establish its millenarian regime. Although the rebellion failed and the dynasty survived, China was still vulnerable for partitioning into spheres of influence under European control. This would have been carried out except for the interference of John Hay, the U.S. Secretary of State, who proposed an

Open Door Policy for trade that eventually was accepted. Yet the integrity of China remained a question throughout the first quarter of the twentieth century.

The height of western influence over the world was reached in the opening years of the twentieth century, but just at this time international tensions over the spheres of influence throughout the various continents were leading up to World War I. The entire human community was so integrated by this time that the concerns of any nation involved the entire complex of nations.

D URING THIS PERIOD when the more comprehensive political contours of the Earth were taking shape, there was also developing inner articulation of the nation communities. This inner articulation under the leadership of the western countries came in the form of nation-states and self-government by the people as the way to enable the human to expand in its personal freedoms and in its cultural forms as well as in its economic power. This emphasis on the individual and the personal rights of the individual belongs to the modern world as one of its more impressive achievements. Yet the harmony between the concerns of the individual and the concerns of the community have never been satisfactorily worked out.

Against the background of the religious concept of the individual soul as immediately created by divine power and as being the supreme value in any consideration, the western world begins with individuals and seeks to establish community by association of individuals. The individual is primary; community is derivative. This is different from the mainly Asian concept of the individual coming into existence within the community. The community is primary; the individual is derivative. Each of these has its inherent bias. Each has its difficulties. Thus we get the difficulties in our quest to reconcile differences between the socialist and the liberal democratic positions. The movement toward popular government, with greater concern for individual freedoms, that developed in the western world both in its ideals and in its institutions, soon spread over the planet. This constitutes one of the main achievements of the modern period. Although throughout these earlier phases women and those without property did not participate in the process, a context was established whereby later liberating developments would become more available for everyone.

In the beginning this liberal democratic commitment came to be accepted only through revolutionary processes, first by the English revolutions throughout the seventeenth century, the American Revolution of 1776, and then the French Revolution of 1789. In each of these the rights of peoples found their first articulation in political form. The American Revolution was the first of the colonial revolutions against an occupying power. The French Revolution was more a social revolution against a ruling class in favor of freedom from oppressive dynastic government.

The nation-state was the main concern in all three instances. For the various peoples to achieve their identity and their freedoms within the democratic setting of the nation has been considered the basic mission of the nineteenth century. Nationalism, together with progress, democratic freedoms, and limitless rights to private property and economic gain, might be considered the pervasive mystique of the late eighteenth, the nineteenth, and the twentieth centuries, first in western Europe and the Americas, then in successive stages throughout much of the human community, a mystique with entrancing benefits but also with devastating possibilities that would only become clear much later when all four of these forces would be recognized as the anthropocentric basis for a temporary human improvement that would eventuate in the ruin that humans would bring upon the natural world. At this time, however, only the bright, the beneficial, the millennial aspects of these movements would be available to human consciousness.

THE SENSE OF BEING a nation provided the integrating community for recent centuries, just as the band was the integrating community in the Paleolithic period, the village the integrating community in the Neolithic period, the capital city and the surrounding supportive territory the integrating community of the literate urban period. The nation in recent centuries has provided the highest appeal in terms of life dedication. National anthems were composed. Flags were designed as the basic symbol of the new community. Banners were a foremost means of self-proclamation during the various dedicated movements that swept over Europe in the course of its history. These were mainly of a religious nature or they proclaimed the power of a ruling dynasty. Since the medieval period heraldry had been a highly developed symbol system indicating a sacred community defiant of the entire outside world.

Although the primary principle of the nation-state is that it recognizes no higher power than itself in the sociopolitical order, by the reference to some sacred symbol in oaths of office the relation of human power with divine authority is still shown to constitute the basis of community rule even in supposedly secular states. The nation provides the unifying constituent reality of the higher self of the individuals composing the community. It quickly established itself as the functional myth of the community. The nation-state can be considered among the most powerful political forms ever invented.

Of primary concern to the nation-state is the territory that it occupies. National boundaries are sacred. To be born within this sacred territory is to be a citizen. The territory must be defended at whatever cost. This contraction of sacred territory from the entire range of the natural world goes with a contraction in the concept of species unity in the human order. The comprehensive sacred community of the entire planet becomes less evident. So too a radical division is created between the citizens of the various nations; where one nation is sacred other nations can be seen as demonic.

Membership in a nation community easily becomes more significant than membership in a religious or cultural tradition. The nation-state has indeed become the sacred community. Although sacred in its political implications, the land occupied is recognized more as territory to be exploited economically than as territory to be communed with spiritually. Or if communed with, this communion belongs to the romantic dimension of life, not to its realistic dimension. In contrast to earlier periods in the Near East where the natural world was addressed as person or as *thou*, the natural world and the land becomes property, an *it*. An economic realism comes on the scene at this same time. The "wealth of nations" becomes a primary focus of concern.

The nation-state was from the beginning an affair of the bourgeois, of those possessing private property. With the emphasis on private property a sense of the individual came into being, and of the right of the individual to own and exploit property, thus the tendency toward individual rights as opposed to community rights. In the Paleolithic period the region had been divided into tribal and hunting territories. In the Neolithic village the land was the village land, the commons, and plots for individual cultivation were identified within this larger context. Ultimately the land belonged to the deity to which the community was dedicated. For the individual, land was less a territory to be owned than an area to be related to for limited purposes.

The educational mission was transferred from its religious-cultural to its secular-social setting. Through this power the nation-state established its historical identity in the populace and communicated to its citizenry the social, political, economic, and cultural ideals of the society. Universal education was begun as the principal instrument for teaching the values of the society. The nation quickly became the primary referent for reality and value.

IN THE CONTEXT of the nation-state wars took on the character of holy wars, whether for liberation of submerged classes into true citizen status, as with the French Revolution, or for liberation from colonial rule, as with the American Revolution, the other revolutionary wars throughout Latin America in the first half of the nineteenth century, and revolutionary wars in Africa in the mid-twentieth century. Then there were civil wars for the unity of the nation, such as took place in America from 1860 until 1865. Dynastic wars to secure unity, independence, and freedom also took place in Italy and Germany in this same period. The Taiping war in China from 1850 until 1865 was anti-dynastic but driven by visionary ideals quite different from the wars associated with national aspirations.

The wars of colonial conquest by nation-states throughout these past two centuries were related to a missionary fervor for the propagation of western bourgeois values to the entire human community, combined with the crasser purpose of economic gain through new markets for the factory production that had been introduced principally in England but later throughout the European world. The exclusive right to trade relations with a given colonial nation was of great economic advantage. The colonies were seen as a market for the new factory productions of the colonizing country, and also as a resource base for unprocessed material. The values were in the woodlands to be cut, the soil for plantation-based money crops, the mines for their ores; all these to be exploited through labor available for minimal recompense. Such were the attractions that drew the colonizing nations to the distant regions of the planet, although greater than any of these attractions was the grasping after power over territory and peoples in a European rivalry to control as much of the planet as possible; to control lest one be controlled.

These colonial wars for western control of non-European lands brought about continual conflict among the western nations themselves throughout the period of colonial expansion. Treaties were constantly being made to partition out the various regions of the Earth in some manner agreeable to these colonial powers. The never-ending strife resulted in World War I and the vain effort afterward to establish an effective League of Nations whereby the peoples of the Earth could live in peaceful relations with each other.

This League of Nations proved most effective in its humanitarian activities but eventually nonviable in the political order. Antagonism between the nations was too intense. The will for peaceful settlement of disputes was almost totally lacking in the member nations. There was too much nationalist resentment between them, too much rivalry for control of territory. Political ideologies seized too violently on Russia, Germany, and Italy. Economic depression settled over the world after 1929. The times were not propitious, though the effort made by the League was admirable.

Just as World War I was ending, in 1917, communism, a political force submerged for decades, erupted in Russia under the leadership of Vladimir Ilyich Lenin. It had such powerful appeal for the oppressed peoples of the world that it quickly spread throughout the world and for seventy-four years remained among the most powerful forces in the human community for challenging colonial powers and the liberal democratic political regimes.

The communist movement had been presented by Karl Marx in the Communist Manifesto written with Friedrich Engels and published in 1848, the most powerful of modern revolutionary statements, challenging at a profound religious, cultural, and moral basis the economic and political structures of the bourgeois social order. The Manifesto combined theoretic power with historical realism and with extraordinary inspiration for organizing the proletariat of the world. Above all it associated the powerful millenarian ideals derived from the biblical Christian tradition with the vision of a classless society to be achieved through the communist movement. This movement itself was considered as the expression of an historical imperative that had governed the course of human affairs from the beginning. To attain an end to bourgeois injustice in the millenial classless society was considered justification for all the agonies experienced in carrying out the social and economic changes indicated.

The tragic course of this movement from its revolutionary inauguration in Russia in 1917 through to its dissolution in 1991 can be accounted as one of the most poignant efforts toward human fulfillment in the course of human history. The millennial dreams of a period of peace, justice, and abundance ended in extreme social turmoil, personal and national impoverishment, and industrial despoliation of the natural environment. At few moments known to us has the inner rage against the human condition hidden deep in the western psyche revealed itself in such extreme historical manifestation. This rage itself evoked excessive dreams of the millennium promises of the scriptures, now to be obtained through an immense effort and even violent taking down of existing social and cultural structures so that the better world might emerge. So much suffering endured, such hopes for a shared community existence. Yet the effort to impose all this violently rather than through peaceful evocation ruined the entire project from the beginning.

The appeal was to those who had not benefited from the bourgeois revolution of the prior century, as well as to those dedicated idealists who were most critical of the exploitive aspects of the democratic institutions as these existed at the time. The need for a radical restructuring of human society in its most basic ideals was evident. The basis proposed by Marx, dialectical materialism, provided a sense that the classless society such as he proposed was a kind of historical inevitability. In this manner he provided a religious fervor for the movement that became a rallying point for immense numbers of people throughout the planet. Thus the twentieth century, in its period from the 1917 revolution in Russia until 1991 when the ideals of Marx as proposed by Lenin and Stalin were rejected, was the time when the communist movement shook the world with its presence.

Yet while all this was happening, another force intervened in human affairs. The National Socialist movement in a Germany victimized by a profound social and cultural pathology threw the entire world into the convulsions of World War II. Eventually the full complex of the nations with their vast capacity for destructive instruments of warfare was drawn into a conflict that raged throughout the Eurasian continent and on the high seas from 1939 until 1945, killing some fifty million people. Consequent on this war also was the rise of nuclear power, first in terms of the atomic and then the hydrogen bomb, and then in 1954 the use of nuclear fission for electrical power.

At the end of this conflict there has arisen a gathering of the nations under the title of the "United Nations," where the destinies of the human community come under discussion and where mediation can take place between the nations with the hope of avoiding wars or at least that the nations involved in conflict situations can meet for discussion of their difficulties before the representatives of the entire community of nations. The effects of modern warfare have become so comprehensive that no society can isolate itself from the effects of such powerful instruments of conflict as are now in use.

POLITICAL IMPERIALISM has given way to economic and cultural imperialism. The northern industrial nations have extended their exploitation of less industrial southern nations in terms of labor, land, and resources. Assistance to the developing nations has resulted in financial servitude, in their growing money crops rather than food crops, and in advantages for large landowners with resources for fertilizers, pesticides, and herbicides and with extensive fields suited for irrigation and for heavy machine cultivation. All of these practices are bringing about further social upset, especially the alienation of women from their traditional roles as keepers of the seeds and caretakers of the soil.

Among the more effective roles played by the United Nations has been to assist in the transition of nations from their colonial status to that of independent nations. That membership in the United Nations has increased from the fifty-one original member nations to one hundred and sixty nations is evidence of the comprehensive nature of its work.

Due to the social and cultural diversity and the economic disparity represented in the nations of the planet, no organization seems to be adequate for bringing about any unity of action except by persuasion, since the nations of the world by definition cannot accept any higher coercive authority. But there is emerging a different kind of authority, something beyond ordinary coercive authority, an authority inherent in the inescapable common destiny of all the nations, a common destiny not only of the human but of every component of the planet Earth.

The human situation in the last decade of the twentieth century is vastly different from that of any previous period in human history or indeed in the history of the Earth itself. The human population of the planet is immensely greater. The power of the human to interfere with the functioning of natural life systems is much greater than it has ever been before. We are also coming to the end of the period of western dominance of the human venture. The order of magnitude of contemporary events involves not simply the human. Present human activities affect all the other components of the planet: the land, the water, the air, and life in all its plant and animal forms.

B<small>ACK OF THESE</small> observations are indications of an even greater need. How to identify and articulate with some clarity a mythic basis adequate for guiding human activity in our present situation? The Paleolithic and the Neolithic were guided by the myth of the Great Mother in their rapport with the forces controlling the natural world. Many tribal peoples identify their rapport with the biosystems of the Earth through totemic symbols. These symbols articulate the intimate relationship that humans have with the various animal and plant forms on which the human depends.

A special need exists for such a mythic basis for the transformation that we are proposing, because the present devastation is the consequence of a powerful myth that has seized the human soul in recent centuries, the myth of Wonderland, the Wonderland that is coming into existence by some inevitability if only we continue on the path of Progress, meaning by Progress the ever-increasing exploitation of the Earth through our amazing technologies. The Wonderland vision is the basis of the advertising that attracts the populace toward ever-increased consumption of products that have been taken violently from the Earth or that react violently with the Earth.

This Wonderland vision is embodied in the Disney World ideal of the human in a nonthreatening world of fabricated imitations, or caricatures of the universe and all its living manifestations. Even further this vision is drawn out to its further possibilities in the Epcot Center associated with Disney World, an artificial world where nothing is left of the original spontaneities of nature. Such is the illusory world that supplies a mythic referent for the present industrialized world and enables society to ignore the manner in which it is irreversibly reducing the planet to Wasteland rather than to Wonderland.

A more valid presentation of the myth that could carry us to a creative renewal of the Earth and all its biosystems we might find in the Universe Story such as it is now available to us. This story, as we now know it for the first time through empirical observation and critical analysis, brings us back to the fifteenth-century renaissance world of intimate presence of all things to each other. It was out of this world of mythic harmonies that the universe was envisaged at the beginning of the scientific age. In virtue of this pervasive harmony each mode of being resonates with every other mode of being. Beings become intelligible precisely through their presence to each other. This sense of an intelligible universe has somehow presided over the entire scientific venture, although only recently have we come to understand just how deeply set the harmonies are in the depths of a world of "chaos." Only in such a context could the mission of mathematics be appreciated as the appropriate instrument for scientific understanding.

Beginning with these harmonies of the universe that could in some manner be expressed in mathematical equations, an intense scientific meditation on the structure and functioning of the universe was begun by western scientists some centuries ago. Among the insights attained by this meditation has been a sense of the curvature of the universe whereby all things are held together in their intimate presence to each other. This bonding is what makes the universe what it is, not a collection of disparate objects but an intimate presence of all things to each other, each thing sustained in its being by everything else.

What we are ultimately talking about in this discussion of the human community in its relation to the natural systems of the Earth is this curvature of the universe whereby things are bonded together in such a way that everything has a shared existence and a common destiny. This universal order that holds things together in a comprehensive embrace can be presented in mathematical equations. But it can be and consistently has been presented in mythic form. From earliest times this vast embrace bonding all things together in the magnificence of the entire created order has been understood in the maternal metaphor as the Great Mother. It was the fecundity and the nurturing quality of the universe that so impressed the earliest humans. This principle of fecundity and this nurturing quality we can now identify with the grand curvature of the universe, for this indeed is the creative and nurturing context of all that exists.

✳ ✳ ✳

Albert Einstein, Pasadena, California, 1931

12.
The Modern Revelation

EVEN WHILE THE nation boundaries and their interrelations were being identified, a momentous change in human consciousness was in process throughout the European world, among the most significant changes since the emergence of human consciousness in the Paleolithic period, a change of such significance in its order of magnitude that we might think of it as revelatory, meaning by this term a new awareness of how the ultimate mysteries of existence are being manifested in the universe about us. In virtue of this revelatory experience we have, over these past few centuries, become aware that the universe has emerged into being through an irreversible sequence of transformations that have, in the larger arc of their movement, enabled the universe to pass from a lesser to a greater complexity in its structure and functioning as well as to a greater variety and intensity in its modes of conscious expression as this can be observed on the planet Earth. This sequence of transformations we might refer to as a time-developmental process.

The change was from a dominant spatial mode of consciousness, where time is perceived as moving in ever-renewing seasonal cycles, to a mode of consciousness whereby the universe is perceived as an irreversible sequence of transformations. This change in perception from an abiding cosmos to an ever-transforming cosmogenesis has had enormous consequences not only in every phase of the human but throughout the entire range of Earth functioning, since this intimate understanding of the universe brought with it almost magic powers of human intrusion on the process through scientific technologies. The failure of humans to understand this change in consciousness, and their corresponding failure to integrate human technologies with the technologies of the natural world, are per-haps the most profound causes for the disturbed condition of the planet in this late twentieth century.

In earlier times the universe and all the beings in the universe were generally perceived as simply being there. The Sun and moon and stars were always in the heavens, so too on Earth the mountains and rivers, the pine trees and the willows and all the birds and animals. Stories of how all this came to be were told in a thousand different ways by the various peo-ples of the Earth, but always in mythic rather than measurable time. The basic movement of the universe was perceived as an ever-renewing seasonal cycle, an eternal cycle, without beginning or ending.

The universe was simply there, an entrancing reality expressive of some ineffable mystery revealed in the awesome, sometimes terrifying reality of natural phenomena. Some infinite energy was at work bringing forth in the annual sequence of nature an endless series of wonders completely over-whelming to the human mind. In some cases, as in India, the universe had come into existence as the day of Brahman, endured for vast periods of time, then dissolved into the night of Brahman, a process that had been repeated forever.

In the classical western world Lucretius referred to this process in his poem *De rerum natura* by saying that *eadem sunt eadem semper*, the same things are ever the same. Within this context Ptolemy had worked out an explanation of how the Sun and moon and Earth and the other stars and planets functioned in relation to each other, an explanation that became the basic context for western cosmology until the time of Copernicus in the early sixteenth century. Religious traditions, taught within the context of the Ptolemaic explanation of the universe, became so intimately related to this explanation that any effort to alter the cosmology was considered as a violation of these most sacred beliefs.

YET EVEN WITHIN this context something new had been added in earlier times by biblical teaching, the sense of historical developmental human time. While the physical universe was fixed in its ever-renewing seasonal sequence, the human component of the universe was in a process of spir-itual transformation in terms of a divine kingdom. Once this spiritual kingdom had attained its full expression within its earthly setting then the physical universe would have served its purpose and would be dissolved while the human would continue in some transcendent realm of blessedness.

Such was the situation in the sixteenth century C.E. when a radical dissatisfaction appeared in western intelligence concerning the universe, its structure and functioning, and the place of the human in the universe. New technologies were emerging, new instruments of navigation, new ways of plotting the seas and continents, new instruments of warfare, new energy sources, new methods of production and commercial distribution. Urban centers were rising, population increasing, university life expanding. Nations were entering into a new sense of their power. Architecture was raising up magnificent structures in a renewal of classical styles. It was a brilliant phase in artistic expression throughout Europe. A new classical learning was spreading over the western world under the title of Latinitas.

Amid all this humans were awakening anew to the natural world, to the structure and functioning and ordering principles of the universe. They were just identifying the continents and seas of the Earth, its plant and animal forms, the variety of its human inhabitants. Others were discovering the laws of motion, the nature of light, the diversity of chemical elements, and their ways of interacting with each other. The ancient mystical experience of the universe was dissolving, the universe that humans had long experienced, communed with, endured and adored, been nurtured and killed by. They had danced to its rhythms, listened to its winds, chanted their exaltation and their grief, wondered at the gorgeous display of the heavens at sunrise and sunset. They had observed the movements of the heavens, identified the planets, marked off the great time divisions, coordinated human celebration with that vast celebration which is the universe itself.

From its earliest times the human community, whether in Paleolithic bands, Neolithic villages, or the monumental cities of the classical period, had experienced these entrancing phenomena but always with such subjective immersion in its mystery as to be unable to step back, observe, describe, and measure with precision the sequence of events and their interrelations. Now there was a change as inquiring minds began to think in quantitative terms, whereas formerly things had been perceived primarily in their qualitative distinction and identity. Much had been achieved in these past periods, but still a closer observation was needed, a more precise description. Above all physical measurements needed to be made and expressed in mathematical terms, for it is through number, as Pythagoras noted, that we understand the nature of things.

For a very long time, certainly since the classical period of Greek thought, little that was new had been learned in the physical, cosmological, or biological realms, few new insights attained. Then suddenly the western mind experienced an urgency, almost a demonic drive toward a more intimate understanding of the realm of matter through more careful observation and experiment, a drive to pass beyond the external barrier of the physical world not into any inner spirit or psychic quality of things but into the inner material forces within things. For this, neither numinous presence nor surface knowledge was sufficient.

A feeling existed also that a more scientific understanding would result in more extensive control over natural phenomena, as though from this early period there existed a dim awareness of the energies hidden deep within the component particles of the universe. An entry into the functioning of the material world had already been made through knowledge of metals. Alchemical procedures were revealing mysterious powers that evoked further inquiry. The sixteenth century was listening to entrancing sounds drawing it into further inquiry. "Curiosity," Aristotle would say. But now apparently something different, a psychic urgency that would not be denied.

Only this could explain the sustained, almost violent assault of western intelligence on the world about it, an assault encouraged in the early decades of the seventeenth century by Francis Bacon, an assault, rather than a communion, that would be sustained over the next four hundred years until eventually it would bring about the transformation of consciousness at its deepest level, a mode of consciousness that would no longer perceive the universe simply as cosmos but as a self-organizing cosmogenesis, a cosmic process expressing itself in a continuing sequence of irreversible transformations.

THIS CHANGE IN consciousness was brought about through a long sequence of advances in scientific understanding from the sixteenth century on into the present. No one knew just what would happen once humans began to look at the universe with a certain direct inquiry into the evidence the universe was giving of itself. The universe of Ptolemy and Aristotle had so controlled European understanding and had entered so profoundly into the structures of western consciousness that to challenge these concepts of the universe was to challenge the entire civilization in its basic sense of reality and value, the foundations not only of personal belief but even of public order.

There was in these centuries an intimate relation between theological and scientific understanding. Newton saw himself as a theologian as well as a scientist. He was describing the manner in which the divine functions in the universe. The same can be said of Leibniz. So with Descartes. For all these a creator-deity was an essential aspect of their cosmological perceptions. It is the basic reason for the difficulty with Galileo, the first person to understand with any clarity the empirical processes of inquiry into physical phenomena.

When Copernicus first broke the spell of the Ptolemaic system, he put into motion a process that neither he nor his successors would understand, a process that would not be understood in any depth until the mid-twentieth century, when an appreciation of the universe as cosmogenesis rather than cosmos came into being. This sense of the universe as self-organizing process was presented in its earliest forms of expression by Henri Bergson, Alfred North Whitehead, Pierre Teilhard de Chardin, and Ilya Prigogine. Even the process thinkers, however, have seldom fully appreciated that the universe in its unfolding is not simply process but a sequence of meaningful irreversible events best understood as narrative.

Even Newton with his understanding of the movement of the heavens had not the slightest idea that the universe held together so marvelously by gravitational attraction was itself in a state of relentless transformation and expansion. He had no idea that there was a story of the universe in the never-ending sequence of its transformations.

To UNDERSTAND JUST what has happened in this five-hundred-year period from the sixteenth through the twentieth centuries, we need to look back to the originators of the scientific mode of inquiry. The first need was to displace a long list of intellectual assumptions: that the celestial bodies were made of material different from the matter of the Earth and follow different physical laws in their movement; that celestial movements must be circular; that the Earth was only some five thousand years old; that the various species of plants and animals were fixed in the beginning; that the universe was best understood as a great chain of being in hierarchical arrangement; that humans were placed on the Earth as a temporary setting for their spiritual development; that the most reliable source of understanding was to follow the teachings of the ancients rather than the observable evidence of the present.

To deal with all these issues required a vast amount of research, the inventions of the telescope and the microscope, as well as the capacity to project hypotheses and then to test these by experiment. Eventually it required the formation of research centers that could focus on problems of enormous complexity with the aid of new mathematical sciences. Technological invention and scientific insight led immediately to other inventions and other insights, so that a vast proliferation of technologies and insights has taken place, at an almost exponential rate of increase.

What needs to be remembered, however, is the larger context of European thought in these centuries, which is found in renaissance learning and in the hermetic mode of thinking. In the esoteric teaching of the period we find the context for the rise of the scientific method of inquiry. The hermetic tradition was especially concerned with the harmonics of the universe, a view of the universe expressed so well in the work of Francesco Giorgi published in 1525, *De harmonia mundi*. Associated with this view of the universe is the role of mathematics in the human understanding of the universe, as established by Pythagoras in the sixth century B.C. (582–507 B.C.). The mathematical mode of modern science owes much to this tradition, which was revived in fifteenth century Italy as a Neopythagoreanism. The positivism of modern centuries is itself rooted in such mythic-mystical traditions. This commitment to the order and intelligibility of the universe has been the basis of the past five hundred years of scientific inquiry. The idea that this intelligibility could be identified in the experiential order was the new sensitivity leading on through the transition phases to our present understanding of the universe. An understanding of this transitional sequence is most important, for only in this way can we appreciate the events that are taking place in these later times. None of the scientists of the seventeenth, eighteenth, or nineteenth centuries knew the larger implications of what they were doing or of the discoveries they were making. Yet each of the major figures was contributing something essential to a pattern of interpretation that would only become clear in the mid-twentieth century. Only now can we see with clarity that we live not so much in a cosmos as in a cosmogenesis, a cosmogenesis best presented in narrative; scientific in its data, mythic in its form.

Most remarkable is the fact that scientific inquiry should for almost three centuries have pursued a course dominated by a mechanistic sense of the universe that would eventually dissolve in the world of relativity theories and quantum physics.

After Copernicus came those archetypal figures mentioned so constantly in any review of this formative period in the story of the modern world: Kepler, Bacon, Descartes, Galileo, Newton. These are especially significant because of their influence on the scientific process itself. The most essential task was establishing physics as the most fundamental of the sciences, the context as well as the model of the empirical sciences. Thus there has been the tendency to reduce even biological studies to molecular physics. Even in the social sciences the effort has been to establish these studies within the norms of reasoning developed in the mathematical and physical sciences.

Johannes Kepler (1571–1630) broke the traditional perception of the circular movement of the celestial bodies by his discovery of the elliptical movement of the planets around the Sun in 1609. Galileo Galilei (1564–1642) established the empirical mode of observation that changed forever the mode of human inquiry into the structure and functioning of the physical world. He was the first to use the telescope effectively in his observations of the heavenly bodies.

René Descartes (1596–1650) discovered analytical geometry and established the mathematical mode of dealing with the physical world. He divided the physical world and the mind into two entirely different realms, with the mathematical manner of association between the two ultimately based on a certain concordism established and maintained by a creator deity. At a single stroke he did away with western consciousness of any inner vital principle of the living world, the sense of soul in the nonhuman

229

world. Until his time, throughout the total course of the various traditions, every living being had had some inner vital principle, an *anima* in the classical world of the west. For Descartes there was no inner principle, no soul. Everything was reduced to matter and its interactions. So too he did away with the sense of inner form as the intelligible principle giving identity to any reality of the phenomenal world. Once this sense of inner form was removed from the material world then the stark subjection of things to mere quantification could take place.

Yet even as we consider this removal of any subjective identity or animating quality of things, we must ask whether this very removal might not have been necessary for a thoroughgoing scientific inquiry into the universe. The consequences of this would later be felt in the larger realm of the human order by displacing those qualities in things that are perceived by the human mind, even the esthetic quality of things. These began to be considered simply as subjective qualities provided by the human mind but not representing any quality found in things themselves. Abandonment of any subjectivity in things, any psychic quality, any esthetic reality, anything beyond quantity, enabled science to penetrate into the functioning of the material world in a manner that might not have been possible if there had been a continued acceptance of the universe as shaped by some inner form or vital principle, or as an expression of some numinous presence, such as had been perceived from Paleolithic times until the rise of this mechanistic science. Severe as this reductionism was, it eventually brought the human mind into a new depth of understanding and bestowed enormous powers of manipulation. That it extinguished the voices of the natural world is also quite true, with consequences that can be observed in succeeding centuries.

A fourth person involved in this re-orientation of human intelligence in its approach to the natural world is Francis Bacon (1561–1626). He established the basic pragmatic orientation that envisaged the scientific venture as serving human welfare, giving to science its commission to besiege the natural world until nature would give up its secrets in the service of the human. This commission of Bacon was so effective because of his enthusiasm for the new scientific methods then beginning to function. His *Advancement of Learning* in 1605 was prior to the work of Galileo in 1610, also prior to Descartes's *Discourse on Method* in 1637. With these three we have the representation of the English, French, and Italian worlds. After Kepler the central European world committed itself to a vitalism and organicism apart from the mechanistic sciences of the French and the English.

I

T WAS THE Englishman Isaac Newton (1642–1727) who gave to the modern world its first comprehensive view of the universe, a view that for over two centuries served as the context for the vast expansion of science that has taken place in this period. Newton gave us an understanding of gravitation as that primary force that holds the universe together in its vast extension in space. Most important, he demonstrated that the laws of gravitation that we experience here on Earth apply to the entire physical world, including those astronomical bodies that we observe in the heavens about us. Once this larger source of universal order in the universe was established, then those persons working in the various other sciences could go about their work with a general feeling of security in what they were about. There were no longer any ultimate mysteries, only the limitation in human efforts to understand.

This began the sequence of research projects that would bring about the Enlightenment movement and the famous French *Encyclopédie* published between 1751 and 1772. Associated with this movement came the sense of progress of the human mind through the centuries. There was a feeling of awakening from the great metaphysical dream world of the past, from unjustified reverence for the earlier sources of human understanding, from the irrational subjection of experimental and observational sciences to the belief systems of the past.

The studies of comparative anatomy in living beings and the efforts at classification during this period established the basic context for the advance of evolutionary understanding in the following century. Efforts toward discovery of the "natural system" in the classification of organisms led Cuvier (1769–1832) rather rapidly to identify four basic classifications: vertebrates, mollusks, articulata (jointed types), and radiata (animals with radial symmetry), forms that he considered were established originally and which endured in various modes of identity. A special field of study from this early period was comparative anatomy, a study of special significance in dealing with fossil evidence in any inquiry into the evolutionary process in its larger time dimensions.

Because our understanding of the life sequence and the geological sequence are mutually dependent, both of these need to be studied together. Only recently have we become aware of just how extensively living organisms have entered into the geological structuring of the planet from an early period, over three billion years ago. There is a basic continuity from this earliest time until the present. At the beginning of modern scientific inquiry into the physical world and its functioning there was a general sense in the western world that the universe had come into being only a few thousand years ago and that it did not really have a history. As an integral reality from the beginning it was sustained by some transearthly power. Only humans had a history: a history within a nonhistorical context.

A new understanding of Earth time began with the work of Georges de Buffon (1707–1788), in his publications in *Natural History,* a series of publications that continued even after his death in 1788. The total of forty-four volumes covered the years from 1749 until 1804. De Buffon was among the first to rethink the age of the Earth, to perceive that the Earth has existed not simply for some five thousand years but for at least eighty thousand years. Even more significant is the work of James Hutton (1726–1797), the Scottish geologist who in 1795 discovered that the geological processes taking place presently can be traced back in time and that an empirically based account of the geological formation of the Earth and of the life sequence can be established. This led to the work of Charles Lyell (1797–1875), who examined the geological structures of Europe and published his description of the structure of the Earth in three volumes in 1830 to 1833.

While these developments were taking place, a parallel series of discoveries was being made in the sequence of organic life as evidenced in the fossils found in the various strata of the Earth. The study of paleontology became our entrance into the transformation sequence of the planetary process. Suddenly, with a shock of recognition, we discovered ourselves as humans emerging within this process, a fact recognized first with some clarity by Jean-Baptiste Lamarck (1744–1829) in his *Zoological Philosophy* (1809), where he traces the evolutionary sequence in a straight line from the lower forms through the higher forms and so on to the human. At least by this time an inner vitalism was asserted in such a way as to overcome the rigid mechanism set forth by Julien La Mettrie in his *Man the Machine* in 1749. This shock of recognition that the human was integral with the natural world was limited in these times to the physical structure of the human and did not significantly affect the basic issue of the human as knowing observer. In this respect the human still remained outside the realm of scientific inquiry. Science was unable to include the knowing subject as simultaneously the object known.

By the time of Lamarck the evolutionary process in the biological order was widely studied and extensively accepted throughout the intellectual life of Europe. Although the evolutionary process remains a mysterious process, understanding was advanced considerably by Charles Darwin (1809–1882), with his theory of natural selection published in 1859. Darwin's theories were taken up and further expanded in their ethical and social implications by Thomas Huxley (1825–1895) and Herbert Spencer (1820–1903). At the close of the century the study of genetics began its most significant development and another element in the evolutionary process, genetic mutation, became a central aspect of the evolutionary explanation leading eventually to a Neodarwinism fostered by George Gaylord Simpson.

As regards theories of the universe during the eighteenth and nineteenth centuries, these had rather limited development after the publication of the *Principia Mathematica* of Isaac Newton in 1687. Only a few efforts were made to deal with the larger astronomical structure and functioning of the universe in these centuries. The basic cosmological structure of the universe was considered to be sufficiently understood. The studies that were done were thought to be minor modifications of the explanations given by Newton. Paul-Henri Holbach (1723–1789) in his *Système de la Nature*, published in 1770, gave the most complete statement of the mechanistic view of the universe. The teachings of Newton were further affirmed in the work of Pierre-Simon Laplace (1749–1827). In his *Treatise on Celestial Mechanics*, published in five volumes from 1799–1825, he committed himself to a totally deterministic position in the functioning of the universe, although his concern was limited to the solar system. Immanuel Kant (1724–1804) in a study of *The General History of Nature and Theory of the Heavens* in 1755 proposed a theory of the formation of the celestial bodies and the solar system through the condensation of gases in the atmosphere.

The true founder of the modern science of astronomy was William Herschel (1738–1822). With powerful telescopes made by himself he surveyed the heavens in a more systematic manner than had been done previously, marking out the various regions and examining them individually.

The other work in the first part of the nineteenth century of general significance in understanding the universe is that of Rudolf Clausius (1822–1888), who with Lord Kelvin (1824–1907) and James Prescott Joule (1818–1889) founded the science of thermodynamics. In 1850 he expounded the Second Law of Thermodynamics, the famous law of entropy, indicating that the amount of energy in the universe remains the same but that the amount of useful energy declines in proportion as the energy is used. This brought up a new science in dealing with both the small- and the large-scale structures and functioning of the universe.

By the opening years of the twentieth century the possibilities for understanding the universe from within the Newtonian perspective were beginning to reach certain limitations. A sense of uneasiness was beginning to appear. The Newtonian universe was too simple a framework. The suppositions underlying the system were too obvious, too little examined, especially the place of the observer in the system. Then, too, problems were beginning to appear that could not be resolved in the Newtonian context, especially the problem of blackbody radiation.

THE NEXT DEVELOPMENT in universe studies altered our understanding of the universe on an order of magnitude surpassing even that brought about by Newton. This transformation was begun primarily by Albert Einstein (1879–1955) in his 1905 papers on Brownian motion, the quantum nature of light, and the special theory of relativity. His work radically altered basic understandings of time, space, motion, matter, and energy. If in the Newtonian world these could be dealt with independently of each other, the first achievement of Einstein was to indicate that they must all be considered in their relation to each other. Especially as regards the equivalence of mass and energy, the simple formula of Einstein that energy equals mass times the speed of light squared has resounded throughout the planet with unlimited consequences both on human understanding and on the affairs of life. The full impact of this simple formula was felt when the inert atoms of Newton suddenly exploded over Hiroshima in 1945.

In a very simple formulation Einstein gave expression to gravitational dynamics as a result of the curvature of space-time. The identification of energy emission in terms of quanta by Max Planck (1858–1947) was further extended by Einstein, who proposed that light consisted of energy particles without electric charge and with wavelike properties, particles that later came to be designated as photons.

In 1900, Planck had originated the theory of quantum mechanics, the theory of energy as emanated in discrete packets called quanta, from his study of radiation from blackbodies when heated. While this required abstruse processes of scientific reasoning, the theory has produced an immense change in our present understanding of the universe and the manner of its functioning both in its microphase and its macrophase reality.

Somewhat later Werner Heisenberg in 1927 produced a radical change in our perception of the universe by his demonstration that completely objective knowledge could not be maintained at the atomic level because the position and the momentum of an electron could not be known at the same time.

Each of these discoveries went deep into the structure and functioning of the universe and our knowledge of the universe. While much of this transformation of our ways of understanding the universe was at the very subtle particle level, another transformation took place through the work of V. M. Slipher (1875–1960) and especially Edwin Powell Hubble (1889–1953), who provided us in 1929 with convincing astronomical evidence that we are living in an expanding universe. Once this had been established and once the background radiation of the universe from its earlier stages had been detected, then all our former knowledge of the universe came together in a remarkable manner.

From the time of Hubble until the closing decade of the twentieth century an immense amount of study of the universe has been carried out. Our studies of astronomy have advanced more rapidly and accumulated more information than in all preceding centuries. We have new instruments of observation. We have plotted the heavens in far more detail than ever before and clarified our knowledge of the planets through instruments sent into space, instruments that have sent back photographs with amazing clarity of detail. With the sensitivity of our electronic instruments we have established contact with a vast range of celestial phenomena. We can outline with some precision the unfolding of the universe during the earliest moments of its origin.

YET THE MORE significant interpretation of it all remained to be made. The single greatest achievement of the entire scientific venture from Copernicus to the present might be a recognition that the data of the universe that we now possess is best understood in terms of a narrative telling of the sequence of transformations that the universe has gone through in its self-shaping emergence throughout space and time. This self-shaping process contains in itself its own directions and its own fulfillment as it has moved over vast periods of time in the larger arc of its process toward a greater differentiation and more intimate bonding, toward a greater variety and intensity in its psychic modes of expression.

That there is a story of the universe in its historical sequence in measurable time was never realized prior to the twentieth century. Nor was this realized even by western scientists throughout the course of their inquiry until very recent times. In the earlier period, especially in the seventeenth century, there was a sense that the human was passing through certain stages in its intellectual development, a sequence later described by Auguste Comte (1798–1857) as the religious, metaphysical, and positivist phases. There was a sense of social transformation toward more acceptable modes of social structure and community life as was articulated by the utopian French socialists, Fourier and St. Simon, also by Robert Owen, and especially by Karl Marx. There was an awareness of the biological development of species from earlier to later forms described by Darwin. There was an outline of the sequence in the geological formation of the Earth given by Lyell.

Yet neither the intellectual development proposed by Descartes and Bacon, nor the social development proposed by Marx, nor the geological

development of the Earth proposed by Lyell, nor the biological development proposed by Darwin, gave any indication that the universe was itself
evolving in an identifiable sequence of irreversible transformations. In
every instance the supposition was that the universe itself was there in
some stable manner.

AT THE PRESENT TIME we know the story of the universe in fragments
rather than in its integrity. Whole libraries are being created with these
fragments: photographs, research papers, plans for further inquiry. The
physical facts themselves are so fascinating that need for further understanding hardly seems appropriate. Both a competence and a willingness
to engage in the immense effort needed to tell the story is what is now
needed, especially if this story is to become what it should be: the comprehensive context of our human understanding of ourselves. This is a task
that requires imaginative power as well as intellectual understanding. It
requires also that we return to the mythic origins of the scientific venture.

Through the renaissance traditions at the origins of modern science we
are able to find our way back to the mythic world of classical times and
further back into the Neolithic and even back into Paleolithic times when
there was an immediate experience of the great liturgy of the universe itself. Only now we have a new understanding of the sequence of transformations that have been taking place over these past eras to shape the
galaxies, to fashion the elements, to gather the solar system together with
its array of planets, to churn together those awesome materials that make
up the Earth, the inner and outer core, the mantle, the asthenosphere, the
lithosphere, the upper crust through the eruptions over the Earth, the violent storms that passed over the planet in its earlier period; then too the
shaping of the seas and continents, the atmosphere and its oxygen, the
coming of life in all the diversity of its plant and animal forms from
the simplest virus to the most elaborate expression, then the emergence of
the human and the sequence of human developments across the planet.
Even our most recent modes of scientific understanding of this immense
story are themselves the latest phase of the story. It is the story become
conscious of itself in human intelligence.

Such reflections as these bring us deep into the realm of imaginative vision where we feel the scientist must participate to some extent in the shamanic powers so charactistic of human presence to the universe in any significant manner. The capacity of Einstein to transform the Newtonian science of his day through his teaching of relativity required a shamanic quality of imagination as well as exceptional intellectual subtlety. So we might say that the next phase of scientific development will require above all the insight of shamanic powers, for only with these powers can the story of the universe be told in the true depth of its meaning.

Such a story was never known before in the course of human affairs. It compares only with those revelatory narratives on which the various cultures of the world were founded in past ages. This story will be dealt with at this level or it will be trivialized and the Earth will for the first time be without a narrative capable of interpreting the historical situation with the grandeur needed.

Just as the universe story has never before been told in this manner, so too the sense of meaning, even the sense of the sacred, that this story carries with it is something new both in its modality and in its order of magnitude. Former stories were mythic narratives dealing with an abiding pattern in the structure and functioning of the universe. This story incorporates the human into the irreversible historical sequence of universe transformations. The important thing to appreciate is that the story as told here is not the story of a mechanistic, essentially meaningless universe but the story of a universe that has from the beginning has its mysterious self-organizing power that, if experienced in any serious manner, must evoke an even greater sense of awe than that evoked in earlier times at the experience of the dawn breaking over the horizon, the lightning storms crashing over the hills, or the night sounds of the tropical rainforests, for it is out of this story that all of these phenomena have emerged. Nor is it the case that this story suppresses the other stories that have over the millennia guided and energized the human venture. It is rather a case of providing a more comprehensive context in which all these earlier stories discover in themselves a new validity and a more expansive role.

* * *

*Women's Resistance to Deforestation, Chipko Movement,
Reni, Uttar Khard Region, Himalayas, 1974*

13.
The Ecozoic Era

T HE UNIVERSE STORY we have traced from its original Flaring Forth through the shaping of the galaxies, the elements, the Earth, its living forms, the human mode of being; then on through the course of human affairs during these past centuries. We could not now leave the reader of this narrative without identifying how this account of the past provides a response to the present and guidance for the future.

Only a mythic vision can do this, since the universe itself can only be presented in a story with a mythic as well as a scientific aspect. Science deals with objects. Story deals with subjects. Since every form of being has both objective and subjective modes, neither is complete without the other.

The terminal phase of the Cenozoic was caused by a distorted aspect of the myth of progress. Though this myth has a positive aspect in the new understanding that we now have of an evolutionary universe, it has been used in a devastating manner in plundering the Earth's resources and disrupting the basic functioning of the life systems of the planet. During this past century humans have taken extensive control over the Earth process with little sensitivity to the more integral dynamics of planetary affairs.

The inner spontaneities guiding the destinies of the natural world have been considered irrelevant. An overlay of mechanistic patterns has been imposed on the biological functioning of the living world. The natural world has become a "resource" for human utilization. Progress has been measured, not by the integral functioning and florescence of the Earth community, but by the extent of human control over the nonhuman world and the apparent benefits that emerged for humans.

That human well-being could be achieved by diminishing the well-being of the Earth, that a rising Gross Domestic Product could ignore the declining Gross Earth Product; this was the basic flaw in this Wonderland myth.

THIS TERM *progress* is itself something of a parody of the inner dynamics of the universe. Progress toward Wonderland, to be achieved by industrial assault upon the planet, is ultimately a subversion of the emergent process whereby the universe in all its aspects has come into being. That the universe, in the diversity and abundance of its expression, has been so successful over vast periods of time is a wonder that we only now begin to appreciate. There have indeed been moments of destruction whereby larger patterns of existence could emerge into being, moments such as happen in the supernova explosions. Yet in these vast transformation events the future possibilities of the entire universe took shape.

This is quite different from the commercial-industrial "progress" that we have witnessed in recent centuries whereby the human has devastated large areas of natural splendor for supposed human benefit. That these centuries of "progress" should now be ending in increasing stress for the human is a final evidence that what humans do to the outer world they do to their own interior world. As the natural world recedes in its diversity and abundance, so the human finds itself impoverished in its economic resources, in its imaginative powers, in its human sensitivities, and in significant aspects of its intellectual intuitions.

We are progressively alienated from any depth experience of the mysterious forces at work throughout the universe. The human soul shrivels into its own being and loses its life-giving contact with all those invigorating experiences of natural phenomena that have guided and energized human activities over the centuries. Communion with these forces through the story of the universe and ceremonial interaction with the various natural phenomena was the traditional way of activating the larger dimensions of our own human mode of being.

Presently we seek to remedy the devastation of the planet by entry into a new period of creativity participated in by the entire Earth community. This new period we identify as the Ecozoic era, a fourth biological era to succeed the Paleozoic, the Mesozoic, and the Cenozoic. These last three terms are conceptual expressions invented in the nineteenth century that enable us to think about the larger patterns of functioning of the bio-systems of the planet. They are subjective, "mythic," organizational expressions with a basis in the observable world although they are themselves not found in the observable world. Since we are terminating the last of these periods, there is need for a fourth term in the same line of expression. This fourth term we designate as the emerging Ecozoic.

THAT THE UNIVERSE is a communion of subjects rather than a collection of objects is the central commitment of the Ecozoic. Existence itself is derived from and sustained by this intimacy of each being with every other being of the universe. We might even suggest that the Earth functions as an organism, provided we understand that we are here using the term *organism* as an analogous expression since the Earth is not simply an enlarged organism at the same level as a tree or a bird. Yet there are similarities between the unity of Earth's functioning and the unity of functioning of any other living being that justifies the use of the term *organic* to describe the inner coherence and integral functioning of the planet Earth. Indeed the Earth is so integral in the unity of its functioning that every aspect of the Earth is affected by what happens to any component member of the community.

Because of this organic quality, Earth cannot survive in fragments. This is one of the most significant aspects of the emerging Ecozoic era. The integral functioning of the planet must be preserved. The well-being of the planet is a condition for the well-being of any of the component members of the planetary community. To preserve the economic viability of the planet must be the first law of economics. To preserve the health of the planet must be the first commitment of the medical profession. To preserve the natural world as the primary revelation of the divine must be the basic concern of religion. To think that the human can benefit by a deleterious exploitation of any phase of the structure or functioning of the Earth is an absurdity. The well-being of the Earth is primary. Human well-being is derivative.

While it is true that the various members of the natural community nourish each other, and that the death of one is the life of the other, this is not ultimately an enmity, it is an intimacy. The total balance in this process is preserved. If there is a taking there is a giving. Without reciprocity the Earth could not survive. Failure to understand this is one of the reasons for the devastation of the late Cenozoic era by its human component. We thought that we could be nourished by the soil without in turn nourishing the soil in accord with its own organic processes and its own rhythms of renewal. We thought that we could exploit the petroleum in the oil fields of the world through our petrochemical industry when we had no way of replacing the carbon in some contained area so that the chemistry of the air could be worked out for the proper functioning of the climate and for the benefit of the life systems of the planet.

This lack of reciprocity has been due to the treatment of the nonhuman world as object for exploitation rather than as subject to be communed with. Especially with the soil: there must be a human communion with the life principles in the soil if there is to be any ultimate benefit for either the soil or the human. The soil is a magic place where the alchemy takes place that enables living forms to survive. So too there is the intimate rapport that humans have with the animal world as this comes down to us from the tribal peoples with their totemic traditions of animals as our venerated ancestors.

This feeling for subjective communion with the various components of the Earth community has been known since the beginning of the human story. Even Paleolithic humans knew this. That is why they did their drawings and paintings of animals, especially those on whom they were dependent. The feeling as well as the esthetic quality of these paintings assures us that the entire world of that period was primarily a communing experience.

If there was also a useful dimension of things, then it had to be mutual. Sometimes the taking and the giving would not be in the same order or by the same individuals; but the circle had to be completed. This was assured by the mutual reverence evoked in the presence of beings to each other and by their mutual dependence for survival.

This intimacy of relatedness extended beyond the living world to the various natural phenomena whereby the universe functions, especially to the sequence of the seasons, to the rain and the wind, to the thunder and lightning and surging of the sea, to the stars and all the other heavenly beings. Everything existed within the single embrace of this immense world where the primordial mysteries of existence were shared in common. In the early civilizations the cosmological order was consistently experienced in terms of human society, and human social order was conceived in terms of the cosmological order. These were different aspects of a single universal order of things.

WHEN WE PROPOSE that the future might be designated as the Ecozoic era we have in mind the restoration, in a new context, of this primordial mode of human awareness. This new context is provided by our more recent experience of the universe as an emergent sequence of irreversible transformations. If, for these earlier peoples, the universe moved simply in ever-renewing seasonal or celestial cycles, the universe in our experience has not only cyclical modes of functioning but also irreversible sequential modes of transformation. These are two different but complementary modes of understanding that cannot be collapsed into a single explanation any more than the wave and particle theories of the phenomenon of light can be reduced to a single explanation.

Moving from our earlier appreciation of the movement of the universe in ever-renewing seasonal cycles to an appreciation of the universe as a sequence of irreversible transformations is, however, a difficulty that seems almost insurmountable by those most committed to this earlier sense of how the universe comes into being and how it functions in its larger patterns of expression.

These two modes of understanding might be compared with the eyesight of an infant. In awakening to the universe, the infant apparently perceives things as extension without depth perspective. Only after a period of time does this other dimension come into infant awareness. So it seems we have, in an earlier moment in history, a certain spatial perspective in our perception of the universe. In this context time moves in ever-renewing seasonal cycles. What comes into being passes away only to be reborn in an everlasting sequence such that as Lucretius tells us, The same things are ever the same.

In this mode of awareness the transformational sequence of an emergent universe is not available to us. The difficulty is enormously compounded by the fact that those who first perceived the universe in its sequence of transformations presented this discovery to us as purely mechanistic, as coming about by meaningless, even by purely random processes. The universe, the stars in the night sky, the dawn and sunset, the immense expanse of the oceans, the mountains seen from a distance, a meadow full of flowers, the soaring of a hawk in the winds of autumn; all this was suddenly reduced to the purely chance association of an infinite number of atomic and subatomic particles.

There is little wonder that this view evoked a revulsion deep in the human soul that has not yet been able to see that this new story of an emergent universe, properly understood as having a psychic-spirit dimension from the beginning, is really an enhancement of all that humans have ever experienced previously in their perceptions of the universe. We now have the wonder, not merely that we are related to and intimate with everything about us, but that we have a cousin relationship with every being in the universe, especially with the living beings of the planet Earth. We have not descended to a lower level; they have, as it were, been recognized at a higher level. Both their lives and ours are infinitely expanded by this intimate presence to each other.

ONE IMPORTANT ASPECT of this new view of the universe is our new realization that the Earth is a one-time endowment. It is indeed an ever-renewing planet, but within limits. Just what these limits are we do not know. But whatever the limits of this planet, it is infinitely precious. No other such planet exists in the solar system. We know of no other such planet in the universe.

The pathos is that we, even now, are deliberately terminating the most awesome splendor that the planet has yet attained. We are extinguishing the rainforests, the most luxuriant life system of the entire planet, at the rate of an acre each second of each day. Each year we are destroying a rainforest area the size of Oklahoma. Not only here but throughout the planet we are not only extinguishing present forms of life, we are eliminating the very conditions for the renewal of life in some of its more elaborate forms.

We have moved from such evils as suicide, homocide, and genocide, to biocide and geocide, the killing of the life systems of the planet and the severe degradation if not the killing of the planet itself. We have moved from simple physical assault on the planet, to disturbance of the chemical balance of the planet through our petrochemical industries, to questionable manipulation of the genetic constitution of the living beings of the planet by our genetic engineering, to the radioactive wasting of the planet through our nuclear industries.

In the terms of the biologist E. O. Wilson, we seem to be bringing about "the greatest impasse to the abundance and diversity of life on Earth" since the earliest beginnings of life some four billion years ago. In the estimate of Paul Ehrlich, another eminent biologist, we are probably extinguishing some ten thousand species each year. Yet another distinguished biologist, Peter Raven, indicates that we may be killing our world.

It has been difficult for humans to appreciate that the planet is given to us as a one-time endowment. Although the Earth is resilient and has extensive powers of renewal, it also has a finite and a nonrenewable aspect. Even the seasonal renewal aspects of the planet can be extinguished: species of plants and animals, for instance. Once a species is extinguished we know of no power in heaven or on Earth that can bring about a revival. The law of entropy is a formidable law.

Iт is already clear that in the future the Earth will function differently than it has functioned in the past. In the future the entire complex of life systems of the planet will be influenced by the human in a comprehensive manner. If the emergence of the Cenozoic in all its brilliance was independent of any human influence, almost every phase of the Ecozoic will involve the human. While the human cannot make a blade of grass, there is liable not to be a blade of grass unless it is accepted, protected, and fostered by the human. These are the three controlling terms in human-earth relations for the foreseeable future: acceptance, protection, fostering.

Acceptance of the given order of things is a more complex affair for humans than it is for other components of the life community since the human has, apparently, a greater capacity for critical reflection on the universe, the Earth, its proper mode of functioning, and our own human role in this larger context of things. Yet for the human, as for every other being, existence is not something acquired or merited by any positive act on the part either of the species or the individual. It is a pure receiving that brings with it conditions determined by the larger course of earthly events, not by the subject brought into existence. This involves acceptance of existence by the species, by the society, and by the individual within the larger context into which it is born.

For the human, existence brings with it immense possibilities of delight and yet a series of never-ceasing anxieties concerning survival, both physical and psychic. Despite the satisfaction experienced in a breath of air, a drink of water, gazing at the stars in the darkness of the night, awakening to the dawn, or communing with the sunset in the evening; despite the joy we have in natural phenomena such as the song of the mockingbird or the taste of a peach; despite the mutual fulfillment in the embrace of child and parent, and humans with each other; despite the folk songs we sing and the music of the great composers, the visual and performing arts, our dances and religious rituals; despite the inner spiritual experiences available to us; we find in the western psyche what might be designated as a deep hidden rage against our human condition. There are, of course, the more ordinary pains to be endured, the heat of summer and the chill of winter, periods of abundance and scarcity, the disappointments in human relationships. There are indeed the distortions, the pains we endure personally and the pain of others. Eventually, we discover, death is the price we pay for life.

For all these reasons there tends to develop a pervasive sense of the pathos of existence. Yet if the delights of existence came with pure, unalloyed joy, with no price to be paid; if death were not the condition of life; then the whole of existence might tend toward the trivial. Danger and death are conditions for adventure; and life without adventure can be a dull experience. Such a mediocre existence cannot be permitted by the very structure and functioning of the universe.

Yet such a mediocre mode of being is precisely what has been invented in the terminal phases of the Cenozoic. With newly acquired power for mechanistic control over the natural world, we have discovered the power to protect ourselves from the elements, to produce food in enormous quantities and transport it anywhere in the world, to communicate instantly throughout the planet, even to delay death by artificial contrivances. With all this knowledge and corresponding skills, we have created a human controlled, less threatening world, a world deprived of the great natural challenges of the past. It also goes with a devastated natural world, for in the process of protecting ourselves from the natural conditions of things, we have done away with many of the most delightful and creative aspects of our existence. What we have gained by controlling the world as a collection of objects we have lost in our capacity for intimacy in the communion of subjects. We no longer hear the voices of all those natural companions of ours throughout the Earth. Nor is it clear that we have really gained the satisfactions we sought.

This new world of automobiles, highways, parking lots, shopping malls, power stations, nuclear-weapons plants, factory farms, chemical plants; this new world of hundred-story buildings, endless traffic, turbulent populations, mega-cities, decaying apartments, has become an affliction perhaps greater than the more natural human condition it seeks to replace. We live in a chemical-saturated world. It is not a life-giving situation. If not deadly it is degraded. Humans now live amid limitless junk beyond any known capacity for creative use. Our vision is impaired by the pollution in the atmosphere. We no longer see the stars with the clarity that once existed.

Yet there has developed a mystique of this sort of existence, especially by those most protected against the full force of the destitution consequent on the total process. There is a feeling of accomplishment, an irrevocable commitment that this is the way into a more creative mode of being for the entire planet since, after all, we are ultimately in control and the planet is here to serve human needs and purposes. Our sciences and our technologies seem fully capable of providing a remedy for any ill consequences of what we are doing. We have extraordinary powers such as those associated with genetic engineering.

A newly developed mystique of our plundering industrial society is committed to moving out of the Cenozoic, not by entry into the Ecozoic, but by shaping an even more controlled order of things that might be designated as the *Technozoic* era. The greater part of contemporary industrial society, it seems, is oriented toward the Technozoic rather than to the Eco-zoic. Certainly the corporation establishment with its enormous economic control over the whole of modern existence is dedicated to the Technozoic.

FROM THIS IT IS clear that a mystique counter to the commercial-industrial mystique must be evoked if the Ecozoic era is to come into being. The future can be described in terms of the tension between these two forces. If the dominant political-social issue of the twentieth century has been between the capitalist and the communist worlds, between democratic freedoms and socialist responsibility, the dominant issue of the immediate future will clearly be the tension between the Entrepreneur and the Ecologist, between those who would continue their plundering, and those who would truly preserve the natural world, between the mechanistic and the organic, between the world as a collection of objects and the world as a communion of subjects, between the anthropocentric and the biocentric norms of reality and value.

Only a comprehensive commitment to the Ecozoic can effectively counter the mystical commitment of our present commercial-industrial establishments to the Technozoic. There is a special need in this transitional phase out of the Cenozoic to awaken a consciousness of the sacred dimension of the Earth. For what is at stake is not simply an economic resource, it is the meaning of existence itself. Ultimately it is the survival of the world of the sacred. Once this is gone the world of meaning truly dissolves into ashes. We will be living in a lunar situation. For in the desolate expanse of the moon our only conception of the divine would reflect the lunar landscape, our imagination would be as bleak as the moon, our sensitivities as dull, our intelligence as blank. We cannot change the outer world without also changing our inner world. A desolate Earth will be reflected in the depths of the human.

The comprehensive objective of the Ecozoic is to assist in establishing a mutually enhancing human presence upon the Earth. This cannot, obviously, be achieved immediately. But if this is not achieved in some manner or within some acceptable limits the human will continue to exist in a progressively degraded mode of being. The degradation both to ourselves and to the planet is the immediate evil that we are dealing with. The enhancement or the degradation will be a shared experience. We have a common destiny. Not simply a common human destiny, but a common destiny for *all* the components of the planetary community.

The immediate goal of the Ecozoic is not simply to diminish the devastation of the planet that is taking place at present. It is rather to alter the mode of consciousness that is responsible for such deadly activities. When we fail to recognize the primary basis for survival in uncontaminated air, water, and soil, and the integral community of life systems of the planet; when we insist on altering the chemical constitution of the atmosphere; when we begin to affect the beneficent climate and the life-giving sequence of the seasons that has brought about the luxuriance of the Cenozoic period into which we were born; then we must consider that we are into a deep cultural pathology. It is particularly pathetic when we bargain over these issues of life and survival for monetary gain or some commercial advantage for a few individuals or a corporative enterprise.

The basic obligation of any historical moment is to continue the integrity of that creative process whence the universe derives, sustains itself, and continues its sequence of transformations. To alter this in some significant manner or to seek to control this process under the assumption that we know how the planet functions in its comprehensive context is an ultimate folly. This sequence of transformations is too mysterious for comprehensive understanding by ourselves or by any form of consciousness that we are acquainted with.

The human mode of understanding, however, does bring with it a unique responsibility for entering into this creative process. While we do not have a comprehensive knowledge of the origin or destiny of the universe or even of any particular phase of the universe, we do have a capacity for understanding and responding to the story that the universe tells of itself, how it emerged in the beginning, the sequence of transformations leading to the wondrous world spread out before us in the heavens and the vast spectacle presented to us by the Earth in its geological, biological, and human manifestations.

As with all other earthly beings, we are expected to enter into this process within those distinctive capacities for human understanding and appreciation that provide our human identity. We are expected to enter into the process, to honor the process, to accept the process as a sacred context for existence and meaning, not to violently seize upon the process or attempt to control it to the detriment of the process itself in its major modes of expression.

THE QUESTION ARISES of a pervasive human responsibility at every level of the human from the individual person through all manner of social and political communities throughout every profession and human association, since our responsibility is to life itself, to the planet Earth, and to the role assigned our existence as a species. Our responsibility now at the end of the twentieth century and the beginning of the twenty-first century can only be fulfilled by assisting in the emergence of the Ecozoic Era that is coming into existence out of the ruins of the Cenozoic.

By entering in to the control of the planet through our sciences and our technologies in these past two centuries, we have assumed responsibilities beyond anything that we are capable of carrying out with any assured success. But now that we have inserted ourselves so extensively into the functioning of the ecosystems of the Earth, we cannot simply withdraw and leave the planet and all its life systems to themselves in coping with the poisoning and the other devastation that we have wrought.

The primary need is to withdraw from our efforts to impose a mechanistic overlay on the bio-systems of the planet, to step back from what we have been doing. Then we might listen to the natural world with an attunement that goes beyond our scientific perceptions. However helpful these may be, they cannot deal with those spontaneities that ultimately determine the course of things in the bio-systems of the planet. In understanding this process something more than science is needed, or a new mode of science such as is suggested by the title *A Feeling for the Organism*—a biography of the biologist Barbara McClintoch.

Our proposal then is not that we walk away from the natural world, but that we follow our own instinctive sensitivities in relating to the natural world. That humans have not, in fact, abandoned the Earth is increasingly evident in the multitude of movements throughout the various regions of the Earth. The orientation we have outlined here is in general the commitment of all serious ecology movements throughout the human community. A multitude of such movements in our social institutions is already functioning.

A deep consciousness of our present plight began with the essay of Aldo Leopold entitled "A Land Ethic" in his *Sand County Almanac,* published in 1948. In that same year Fairfield Osborn published his book entitled *Our Plundered Planet.* Both of these were some years before the effects of the DDT assault on the biosystems of the planet had evoked the response of Rachel Carson in *Silent Spring,* published in 1962. A furor erupted from both scientific and commercial communities. The petrochemical industry was just beginning its expansion at this time, with little concern for the impact that these thousands of new chemicals each year were having on the functioning of the planet.

As the pollution of the water and the air increased throughout this period, there developed a growing consciousness that severe difficulties were arising as regards the consequences in terms of human well-being. That this required action on the part of government was sufficiently evident that the Clean Air Act was passed by the Congress of the United States in 1970. Soon it was also clear that the air and the water were constantly in circulation around the globe and that a transnational solution was needed.

Such consideration was begun in a serious way at the Stockholm Conference sponsored by the United Nations in 1972. At the time there were no environmental protection agencies anywhere in the world. Since that meeting most of the nations have established such agencies, but even more significant than these is the United Nations Environmental Program.

In this same year, 1972, an extraordinary work appeared as a report to the Club of Rome, *Limits to Growth,* a first comprehensive presentation out of the new science of systems analysis. This attack on the folly of indefinite progress and the consequent impoverishment of the planet that was taking place evoked furious opposition from both scientific and commercial sources. Many of the critics missed the main point of the study, which was the absurdity of exponential rates of increase in consumption patterns, in financial affairs, in population, or in anything whatever.

In 1980 a World Strategy for Conservation and Development was developed by over seven hundred scientists and other professional representatives from over one hundred nations indicating that there is no way to attain human well-being apart from the well-being of the ecosystems of the Earth.

In 1982 the United Nations approved a World Charter for Nature articulating the intimate relationship of humans with the other living forms and the inherent rights of every living form to its necessary habitat.

In 1987 The World Commission on Environment and Development appointed by the United Nations Assembly in 1983 submitted its report entitled *Our Common Future*. This was an effort to summarize the present status of human-Earth relations with some indication of directions in which the human community should direct its activities in coming years if we were to avoid impending crises beyond human capacity to deal with. The outstanding achievement of this report has been its insistence that there cannot be any solution for the present threats to survival by any one nation for we do not have separate futures; there is only a shared future, a common destiny for the entire company of nations as well as for the larger Earth community itself.

THAT OUR WESTERN civilization should be the principal cause of such extensive damage to the planet is so difficult a truth for us to absorb that our society in general is presently in a state of shock and denial, of disbelief that such can possibly be the real situation. We are unable to move from a conviction that as humans we are the glory and the crown of the Earth community to a realization that we are the most destructive and the most dangerous component of that community. Such denial is the first attitude of persons grasped by any form of addiction. Our western addiction to commercial-industrial progress as our basic referent for reality and value is becoming an all-pervasive attitude throughout the various peoples and cultures of the Earth.

Efforts to present the full reality of the situation are being met generally with intense opposition, an opposition due in large measure to the subservience of our religious, educational, and professional establishments to our industrial culture. These major determinants of our cultural forms are manifesting minimal concern for the catastrophic situation that is before us.

The remedy would seem to emerge, as in denial situations generally, in a crash so severe that we are suddenly confronted with a choice between death and abandoning our addictive mode of functioning. In this case the impending crash is of immensely greater dimensions and more definitive consequences. The crash that faces us is not the crash simply of the human, it's a crash of the bio-systems of the Earth, indeed it is in some manner the crash of the Earth itself.

We do not imply here that the more elementary life systems of the planet are in danger of extinction. These, the microforms of life—along with the insects, rodents, plants, and many of the trees and animals—will surely continue, probably in ever greater abundance. What is indicated here is that the conditions of the planet in any foreseeable future, or in that phase of historical or biological time available to human understanding, would suffer severe deterioration in terms of their present mode of expression. The water and air and soil pollution could remain for a significant period, along with the decline of the rainforests, the diminishment of life species.

Yet the tendency to minimize the difficulties before us is the greatest obstruction to the radical change in human consciousness, a change at the order of magnitude required for entry into the creative phase of the Ecozoic. This change requires something of a different order but equivalent to a new religious tradition. Our new sense of the universe is itself a type of revelatory experience. Presently we are moving beyond any religious expression so far known to the human into a meta-religious age, that seems to be a new comprehensive context for all religions.

THE ECOZOIC ERA requires a comprehensive human consensus. It needs such support for its planetwide programs. The entire planet would then be considered as a commons. Already the atmosphere, the seas, and the space above the Earth are being recognized as areas of universal relevance. There are also biological areas of global concern.

The human professions all need to recognize their prototype and their primary resource in the integral functioning of the Earth community. The natural world itself is the primary economic reality, the primary educator, the primary governance, the primary technologist, the primary healer, the primary presence of the sacred, the primary moral value.

In economics it is clear that our human economy is derivative from the Earth economy. To glory in a rising Gross Domestic Product with an irreversibly declining Gross Earth Product is an economic absurdity. So long as our patterns of consumption overwhelm the upper reaches of Earth's sustainable productivity, we will only drive the Earth community further into ruin. The only viable human economy is one that is integral with the Earth economy.

Education is already late in its revision, but we can expect that it will in the future be extensively altered. Education might well be defined as knowing the story of the universe, of the planet Earth, of life systems, and of consciousness, all as a single story, and recognizing the human role in the story. The primary purpose of education should be to enable individual humans to fulfill their proper role in this larger pattern of meaning. We can understand this role in the Great Story only if we know the story in its full dimensions.

In our governance we are moving from a limited democracy to a more comprehensive biocracy. Already we can envisage a constitution not simply for humans on this continent, but for the entire North American community, including both the geographical structures and functioning of the Earth and the various life systems dispersed over the Earth. A beginning has been made in the legislation requiring environmental impact statements before any major project affecting the environment can be undertaken. Among the challenges that face governance in the human order is the relationship of national governments with each other, with particular reference at the present time to the more industrialized northern nations of the planet that are exploiting the less industrialized southern nations. This currently is not only negating the human advancement but is bringing about a pervasive devastation of the human societies.

The engineering profession begins to learn that human activity is most effective and most enduring when it is in accord with the natural functioning of the ecosystem into which it is inserted. Human engineering needs to be guided by the masterful technologies of Earth. In the transport systems of vascular plants, in the energy conversions of photosynthesis, in the efficiencies of the hydrological and mineral cycles, in the communication systems of the genetic codes are technological models of great power and elegance. Needed now are biocentric human technologies that are coherent with Earth technologies, and that would enable our agricultural, energy, and architectural projects to be carried out in an integral relationship with Earth's functioning.

The medical profession begins to see that the well-being of the ecosystems of the planet is a prior condition for the well-being of the human. We cannot have well humans on a sick planet, not even with all our medical science. So long as we continue to generate more toxins than the planet can absorb and transform, the members of the Earth community become ill. Human health is derivative. Planetary health is primary.

Religion begins to appreciate that the primary sacred community is the universe itself. In a more immediate perspective, the sacred community is the Earth community. The human community becomes sacred through its participation in the larger planetary community.

In morality we are expanding our moral sensitivity beyond suicide, homicide, and genocide to include biocide and geocide, evils that were not recognized in our civilizational traditions until recently.

Beyond all this and in a sense more encompassing than any of these is the role of women in the future. The need presently is recognition of women in their capacity to interpret the human venture at its most basic level in the context of the universe and the planet earth. The family, child-bearing and child-rearing, will always be a central focus of human concerns. Yet for women especially a new range of their activities is needed both for themselves and for the larger destinies of the human-Earth venture. We cannot do without those special insights that women offer in every phase of human existence.

The need to limit population increase and to even consider a possible decrease in population has altered in a definitive manner the entire human situation in one of its most basic aspects. This leads immediately to new possibilities in women's participation in the course of human affairs that has been consistently thwarted in most traditional civilizations. The rise of movements such as feminism and ecofeminism has already altered all the basic professions and social institutions throughout the industrialized countries of the world. As this participation increases throughout the world, as women are liberated from the oppressions they have long endured, as women reach new levels of personal fulfillment, a new energy will undoubtedly be felt throughout the Earth. While we cannot know just what the consequences will be, there is reason to hope that this will be a vast creative and healing energy.

As REGARDS LANGUAGE, we have so far had a human-centered language. We need an Earth-centered language. A new Ecozoic dictionary is gradually taking shape as we begin to articulate our new vision more clearly. All the more substantive words in the language are undergoing a transformation, words such as *society, good* and *evil, freedom, justice, literacy, progress.* All these words need to be extended to include the various beings of the natural world, their freedoms, their rights, their share in the functioning of the Earth.

Our Cenozoic dictionary cannot deal adequately with the realities of existence in this new period. We need an Ecozoic dictionary.

Beyond any formal spoken or written human language are the languages of the multitude of beings, each of which has its own language given to it generally, in the world of the living, by genetic coding. Yet each individual being has extensive creativity in the use of the language. Humans are becoming much more sensitive to the nonhuman languages of the surrounding world. We are learning the mountain language, river language, tree language, the languages of the birds and all the animals and insects, as well as the languages of the stars in the heavens. This capacity for understanding and communicating through these languages, until now enjoyed only by our poets and mystics, is of immense significance since so much of life is lived in association with the other beings in the universe.

Among the greatest changes linguistically is the change from our present efforts at an exclusively univocal, literal, scientific, objective language to a multivalent language much richer in its symbolic and poetic qualities. This is required because of the multivalent aspects of each reality. Scientific language, however useful in scientific investigation, can be harmful to the total human process once it is accepted as the only way to speak about the true reality of things. A more symbolic language is needed to enter into the subjective depths of things, to understand both the qualitative differences and the multivalent aspects of every reality.

Indeed all our basic words are multivalent terms that are used analogously. This simply means that there are qualitative differences in their various meanings. Precisely because of its organic mode of being, the Earth is not a global sameness, Earth is a highly differentiated unity as is every organic reality, so that each component bioregion has its own unique mode of integration and its own special role to fulfill that needs to be recognized and responded to if any well-being is to be achieved by the planet in its integrity.

The sciences and the humanities, business and religion, the arts and sciences; all these divisions of learning are beginning to overcome their isolation from each other. Even though the distinctive roles of each need always to be recognized, they will hopefully, in the future, become much more integrated with each other. All our professions and institutions need to be appreciated in the light of the single story that governs the basic functioning of the Earth as well as the entire human process.

We begin to rethink the structure and functioning of our cities. No longer need we endure without protest the oppression of the automobile as the primary factor in our city architecture. Already cities are being redesigned to bring the streams above ground rather than condemning them to flow through our sewers. Our cities begin to be places for habitation not merely by humans but also by other life forms. We begin to make provision for the birds and the various animals that are proper to the region.

There is no reason why the various life forms, the fish in the streams, the flowering plants, the animals proper to the region, should not share in our common habitat. We diminish the grandeur of human habitat once we exclude other living forms from their sharing in the single life community. The need of things for a distance from each other at times need not indicate that the variety of living forms can be diminished without harming the inherent vigor of the community itself and all its component members.

Earlier we were concerned simply with our own limited area. We withdrew from the major forces of life into the realm of our own limited controls. We developed our individual self with a neglect of our community self, our relation with the planet Earth, and with the entire natural order that constitutes the larger self of our own being.

What we seldom think about is the human as species. We will never come to appreciate the full significance of human adjustment in this new biological era until we begin to think of the human as a species among species. We have thought about the human as nations, as cultures, as ethnic groups, as international organizations, even as the global human community; but none of these articulates the present human-Earth issue so precisely as thinking about the human as a species among species.

We need an inter-species economy, an inter-species well-being, an inter-species education, an inter-species governance, an inter-species religious mode, inter-species ethical norms. Until we begin to think about our human story as integral with the larger life story and the larger Earth story, we will not be fully into the Ecozoic period. We will not have an Ecozoic governance.

A FINAL INTEGRATION of the Ecozoic era with the larger pattern of the Universe Story such as we are narrating it here has to do with the curvature of space-time. The nature of this curvature is bound up with the energy of the universe's primordial Flaring Forth. Had the curvature been a fraction larger the universe would have immediately collapsed down into a massive black hole; had it been a fraction smaller the universe would have exploded into a scattering of lifeless particles. The universe is shaped in its larger dimensions from its earliest instant of emergence. The expansive original energy keeps the universe from collapsing, while the gravitational attraction holds the component parts together and enables the universe to blossom. Thus the curvature of the universe is sufficiently closed to maintain a coherence of its various components and sufficiently open to allow for a continued creativity.

This balance of forces is again what gives to the Earth its special qualities. Its own gravitational attraction holds the Earth together. The resultant inner pressures combined with the Earth's inner nuclear energy keep the Earth in a state of balanced turbulence whence its continued transformation takes place. Because this balanced turbulence was not achieved in the other planets they were unable to bring forth such living forms as emerged upon the Earth.

We might say that the industrial age, from its human influences, has so upset this balance that the planet has gone into a burnout phase of its existence. The industrial plundering of the planet has increased the toxic condition of the Earth to a point beyond what the Earth can manage creatively. The Earth cannot dispose adequately of the chemical residues spewed into the atmosphere by the burning off of the fossil fuels that the Earth had kept stored for millions of years in order to achieve the chemical balance needed for the expansion of the life systems of the planet.

What the Ecozoic era seeks ultimately is to bring the human activities on the Earth into alignment with the other forces functioning throughout the planet so that a creative balance will be achieved. When the curvature of the universe, the curvature of the Earth, and the curvature of the human are once more in their proper relation, then Earth will have arrived at the celebratory experience that is the fulfillment of earthly existence.

Epilogue
Celebration

THE EVOCATION OF THE Ecozoic era requires an entrancement within the world of nature in its awesome presence: whether in the Himalayan mountains or the Maine seacoast, the Pacific islands, the Sahara desert, the Greenland glacier, the Amazon rainforests, the Arctic snowfields, the prairies of mid-America, or any of the other fantastic presentations that nature makes of itself wherever we look. We need the expansive Earth, the distant sky, the flowering plants and trees and all that multitude of living forms about us; butterflies and bluebirds, Siberian tigers and tropical chimpanzees, dolphins and sea otters and the great blue whale. We know of no other place in the universe with such gorgeous self-expression as exists on Earth. This exuberance of life we see especially in the tropical rainforest with its unnumbered species of flowering plants and colorful insects and the full spectrum of living creatures of every kind. While everything exists for everything else we can only surmise the diminishment of the human mind and imagination if we did not have such magnificence in the natural phenomena about us.

Nor is all this a fixed context of existence. If we were to choose a single expression for the universe it might be "celebration," celebration of existence and life and consciousness, also of color and sound but especially in movement, in flight through the air and swimming through the sea, in mating rituals and care of the young. But then too there is the pathos of both living and dying, of consuming and being consumed. There is the vast hydrological cycle with its sequence of abundance and scarcity, its expression of the tragic as well as the delightful moments of temporal existence.

We belong to this community of Earth and share in its spectacular self-expression. This is the setting that seems to be implicit in the movements toward ecological integrity in this late twentieth century. It is also the ideal that we find articulated in the earliest efforts of the various tribal cultures as well as in the earliest efforts toward the more complex cultures. While the human situation is definitively changed from this earlier period we remain genetically coded toward a mutually enhancing presence to the life community that surrounds us.

The universe we might consider as a single, multiform, sequential, celebratory event, as is implied in the designation of a flock of larks as an exaltation of larks, a title with implications of flight and song expressing delight in existence. For even the afflictions endured cannot diminish the songs that resonate throughout the natural world.

So the universe as a community of diverse components rings with a certain exultation and joy in being, while experiencing the sacrificial dimension of a natural world that tears at every leaf on a tree, that enables only a few seeds to mature out of millions cast abroad. The emergent universe can be considered as a continued elaboration of this sequence of existence and extinction, the pressing toward expanded modes of being and ever more intimate presence of things to each other. Everything about us seems to be absorbed into a vast celebratory experience. Whatever be the more practical purposes of existence it appears that celebration is omnipresent, not simply in the individual modes of its expression but in the grandeur of the entire cosmic process. This larger perspective is often referred to as the cosmic liturgy insofar as it expresses the awesome qualities of phenomenal existence.

This awesome aspect of the universe is found in qualitatively different modes of expression throughout the entire cosmic order but especially on the planet Earth. There is no being that does not participate in this experience and mirror it forth in some way unique to itself and yet in a bonded relationship with the more comprehensive unity of the universe itself. Within this context of celebration we find ourselves, the human component of this celebratory community. Our own special role is to enable this entire community to reflect on and to celebrate itself and its deepest mystery in a special mode of conscious self-awareness.

In the various stages of the human, from its earlier tribal forms throughout the shaping of the more complex civilizations, this celebratory experience is consistently associated with the sense of the sacred. This we find with the aboriginal tribal peoples of Australia as well as with the more complex civilizations of the Eurasian and American worlds. The Native American peoples were especially distinguished for their sense of participating in a single community with the entire range of beings in the natural world about them. The drumbeat was experienced as striking out the rhythm of the Earth itself. Their dancing was associated with the various animal peoples inhabiting the area. It was all caught up in the sacred realm of the Manitou, Orenda, or Wakan Tanka. This we find especially in the ecstatic experience of Black Elk where at the song of the visionary stallion, "The virgins danced, and all the circled horses, the leaves on the trees, the grasses and the hills and in the valleys, the waters in the creeks and in the rivers and the lakes, the four-legged and the two-legged and the wings of the air—all danced together to the music of the stallion's song."

Most brilliantly in the higher civilizations, we find in the Chinese world an elaborate system of coordination with the rhythmic transformations observed in natural phenomena. In the Chinese *Book of Ritual* we find that the emperor himself and the court in its entire functioning are involved in the integration of human affairs with the surrounding world. The music of the court, the colors of garments, the place of residence, the emotional mood, everything must be coordinated with the cosmological structure of the universe and the seasonal sequence of its transformations. Only in this context was there any effective authority among humans, any social order, any artistic creativity. The high expression of this intimacy of the human with the natural world is found in Chinese landscape painting of the twelfth-century Sung Dynasty.

In architecture especially human structures have consistently been aligned with the mythic forces of the universe in their physical manifestation. The orienting point is taken from the point of the rising sun, and the various directions of north and south, east and west are not simply complementary differences; they are rather qualitatively different. These heavenly forces of the stars are understood as determining human affairs in a very direct manner.

265

In their alignment with the stars the people of India are especially observant of the propitious times for the proper arrangement of significant human affairs. In the great altars of the Ashmaveda sacrifice the entire universe is depicted as the proper setting for the consecration of the king as *cakravartin,* as someone binding together the cosmological and the human realm s into a functional unity.

In the Buddhist world the universe exists in and is sustained by the pervasive reality of the Buddha nature, which is present throughout the entire cosmological order. In the Kegon tradition of Buddhism particularly this cosmological presence is discerned with special clarity as the context in which the human finds its identity and fulfillment.

In Dionysius the Aereopagite, the culminating Neoplatonism of Greek philosophy and the Christian mystical tradition join in the teaching that because of their unity of origin all things are bound together in the intimacy of "friendship," the intimacy that justifies the use of the word "universe" to indicate that the diversity of things exists not in separation but in a comprehensive unity whereby all things are bonded together in inseparable and everlasting unity.

THE HISTORICAL ROLE of the present might be to achieve an integration of the human with these natural forces and see in this alignment the type of realization toward which the human might move presently. The most consistent expression of this alignment is found perhaps in the renewal celebrations that originated deep in the Paleolithic period and have continued on through the Neolithic and into the classical civilizations, renewal ceremonies that still take place in many parts of the Earth, most often in alliance with the springtime fluorescence that occurs especially in the temperate climates of the natural world.

One of the differences between the present and the earlier liturgies uniting human activities with the cosmological process is the need for celebration of the great historical moments in the unfolding of the universe. So far events of such immense significance as the supernova implosions have not been appreciated as moments to be celebrated. Yet these implosions of the first generation of stars are what brought about the pattern of elements that are the conditions for emergence of a planet such as the Earth and its life systems. These explosive moments constitute a psychic-spiritual transformation as well as a moment of physical transformation. Only when we appreciate the full significance of this moment can we tell the story of the universe in some adequate fashion. This emergence of more complex structure in the atomic components of the universe is also the period of higher differentiation throughout the entire structure of the universe. As with all originating moments, so this moment has its sacred aspect.

Another moment of vast significance is the moment when photosynthesis began to function some three billion years ago, the period when cells appeared with the ability to turn the energy of the sunlight into substance capable of supporting organic life. This released energy once again of a psychic-physical nature.

After photosynthesis, one of the great moments is the emergence of trees, first the conifers but then the broadleaf trees with their special powers for utilizing the energy of the Sun and providing the wood that was needed for the further advance of life.

Then came the flowers some one hundred million years ago, when blossoms and seeds first appeared with their concentrations of protein that made energy available in small bundles that could be consumed quickly that animals would not need to spend an excessive amount of time eating to sustain life. Beyond this utility aspect of the flower revolution there is the remarkable psychic advance that deserves periodic celebration.

Such advance in the living world needs to be integral with the type of celebration proper to humans, since each of these transformations affects the structure and functioning of the human in both its psychic and physical aspects. The human exists only within this transformation sequence.

It is the special capacity of the human to enable the universe and the planet Earth to reflect on and to celebrate, not simply the present moment, but the total historical process that enables this moment to be what it is.

If we can appreciate the flight of the birds through the sky, how much more profound is it to witness and to celebrate the transformation moment when the flight and song of the birds first appeared. Through our human awareness the entire universe in all its various manifestations comes to itself with a special delight. This capacity to celebrate in our music and our art, our dance and our poetry, and in our religious rituals, is what traditionally has provided the highest forms of human fulfillment.

Until the present we have not been able to celebrate properly this larger story of the universe, yet this is the high achievement of our scientific inquiry into the universe. Once we begin to celebrate the story of the universe we will understand the attraction that so draws our scientists to their work, why every detail of our scientific inquiry becomes so important.

Without entrancement within this new context of existence it is unlikely that the human community will have the psychic energy needed for the renewal of the Earth.

This entrancement comes from the immediate communion of the human with the natural world, a capacity to appreciate the ultimate subjectivity and spontaneities within every form of natural being. We are discovering anew our human capacity for entering into the larger community of life, something that we have not experienced in any adequate manner since our Neolithic origins. This new experience enables us to activate the more extensive dimensions of our own being. Indeed our individual being apart from the wider community of being is emptiness. Our individual self finds its most complete realization within our family self, our community self, our species self, our earthly self, and eventually our universe self.

There is eventually only one story, the story of the universe. Every form of being is integral with this comprehensive story. Nothing is itself without everything else. Each member of the Earth community has its own proper role within the entire sequence of transformations that have given shape and identity to everything that exists.

Timeline

Note: All dates should be taken as approximations. *Timescale,* by Nigel Calder, is the source for many of the dates.

THE PRIMORDIAL FLARING FORTH

15 billion years ago
The universe begins as stupendous energy
The original universe activity evolves into the gravitational, strong nuclear, weak nuclear, and electromagnetic interactions
Before a millionth of a second has passed, the particles stabilize
The primal nuclei form within the first few minutes
As the universe becomes transparent, hydrogen and helium come forth
The galaxies are seeded

GALAXIES AND SUPERNOVAS

10–14 billion years ago
The universe breaks into galactic clouds
The primal stars appear
The first elements are forged in the stars
The first supernovas give rise to second- and third-generation stars
Giant galaxies evolve by swallowing smaller galaxies

THE SOLAR SYSTEM

5.0 billion years ago	A disclike cloud floats in the Orion arm of the Milky Way Galaxy
4.6 billion years ago	Tiamat goes supernova
4.5 billion years ago	Sun is born
4.45 billion years ago	Planets are formed; Earth brings forth an atmosphere, oceans, and continents
3.0 billion years ago	Moon and Mercury's geological activity is frozen
1.0 billion years ago	Mars's geological activity is frozen

LIVING EARTH
(Archean Eon)

4.0 billion years ago	Aries, the first prokaryotic cell, emerges
3.9 billion years ago	Promethio invents photosynthesis
2.5 billion years ago	Continents stabilize
2.3 billion years ago	First Ice Ages
2.0 billion years ago	Prospero learns to deal with oxygen and proliferates

EUKARYOTES
(Proterozoic Eon)

2.0 billion years ago	Vikengla, the first eukaryotic cell, develops
1.0 billion years ago	Kronos discovers heterotrophy
1.0 billion years ago	Sappho invents sexual reproduction
700 million years ago	Argos, the first multicellular animal, appears
600 million years ago	The Mesocosm: jellyfish, sea pens, flat worms
570 million years ago	Cambrian extinctions: 80–90 percent of species eliminated

PLANTS AND ANIMALS
(Phanerozoic Eon)

PALEOZOIC ERA

Cambrian

550 million years ago   Invention of the shell by trilobites, clams, and snails

Ordovician

510 million years ago   Vertebrate animals
440 million years ago   Ordovician catastrophe

Silurian

425 million years ago   Jawed fishes appear
425 million years ago   Capaneus goes ashore
415 million years ago   Development of the fin

Devonian

395 million years ago   Insects
380 million years ago   Lungs appear in the fish
370 million years ago   Devonian catastrophe; invention of the wood cell by the lycopods; the first trees; vertebrates go ashore; amphibians

Carboniferous

350 million years ago   **Land-worthy seeds by the conifers**
330 million years ago   Wings by the insects
313 million years ago   Reptiles show up, land-worthy eggs

Permian

256 million years ago   Therapsids, warm-blooded reptiles
245 million years ago   Permian extinctions: 75–95 percent of all species are eliminated

MESOZOIC ERA

Triassic

235 million years ago	Dinosaurs appear, flowers spread
220 million years ago	Pangaea (supercontinent) is complete
216 million years ago	First mammals
210 million years ago	Birth of the Atlantic Ocean, the breakup of Pangaea

Jurassic

150 million years ago	Birds

Cretaceous

125 million years ago	Marsupial mammals
114 million years ago	Placental mammals
70 million years ago	Primates on the scene
67 million years ago	Cretaceous extinctions

CENOZOIC ERA

Paleocene

55 million years ago	Rodents, bats, early whales, premonkeys, early horses

Eocene

40 million years ago	Various orders of mammals complete
37 million year ago	Cosmic impact: Eocene catastrophe

Oligocene

36 million years ago	Monkeys
35 million years ago	Early cats and dogs
30 million years ago	First apes
25 million years ago	Whales become largest marine animals of all time; carnivores take to the sea and become seals

Miocene

24 million years ago	Grass spreads across land
20 million years ago	Monkeys and apes split
19 million years ago	Early antelopes
15 million years ago	Cosmic impact: Miocene catastrophe

12 million years ago	Gibbons
11 million years ago	Surge in grazing animals
10 million years ago	Orangutans
9 million years ago	Gorillas
8 million years ago	Modern cats
7 million years ago	Elephants
6 million years ago	Modern dogs

Pliocene

5 million years ago	Chimpanzee, hominids: Australophithecus afarensis
4.5 million years ago	Modern camels, bears, and pigs
4.0 million years ago	Baboons
3.7 million years ago	Modern horses
3.5 million years ago	Early cattle
3.3 million years ago	Current Ice Ages begin
2.6 million years ago	First humans: Homo habilis
1.8 million years ago	Modern big cats, bison, sheep, wild hogs

Pleistocene

1.5 million years ago	Hunters: Homo erectus
1.0 million years ago	Mammalian peak
730,000 years ago	Cosmic impact: Pleistocene catastrophe
700,000 years ago	Brown bears
650,000 years ago	Wolves
500,000 years ago	Llamas
300,000 years ago	Archaic Homo sapiens
200,000 years ago	Cave bears, goats, modern cattle
150,000 years ago	Wooly mammoth
120,000 years ago	Wildcats
72,000 years ago	Polar bears

HUMAN EMERGENCE

Lower Paleolithic

2.6 million years ago	First humans, Homo habilis, stone tools
1.5 million years ago	Homo erectus, hunting
500,000 years ago	Clothing, shelter, fire, hand axes
200,000 years ago	Archaic Homo sapiens

Middle Paleolithic

    100,000 years ago   Ritual burials

Upper Paleolithic

    40,000 years ago   Modern Homo sapiens; language; occupation of
                                   Australia

    35,000 years ago   Occupation of the Americas

Aurignacian

    32,000 years ago   Musical instruments

Gravettian

    20,000 years ago   Spears, bows and arrows

Magdalenian

    18,000 years ago   Cave paintings

THE NEOLITHIC VILLAGE

12,000 years ago	Dogs tamed
10,700 years ago	Sheep and goats tamed in Middle East
10,600 years ago	Settlements in the Middle East; wheat and barley cultivated in Middle East
10,000 years ago	Dogs tamed in North America
9,000 years ago	Settlements in Southeast Asia: rice gardeners; water buffalo, pigs, and chickens tamed. Painted pottery culture
8,800 years ago	Cattle tamed in Middle East
8,500 years ago	Settlements in the Americas: cultivation of corn, squash, peppers, and beans; weaving in Middle East
8,000 years ago	Irrigation in Middle East; population of Jericho is 2,000
7,500 years ago	Hassuna culture; millet farmers in North China
7,000 years ago	Çatal Hüyük population is 5,000
6,400 years ago	Horses tamed in Eastern Europe
5,300 years ago	Pottery in the Andes

5,000 years ago	Early European settlements; gourds, squash, cotton, amaranth, and quinoa in the Andes; camels and donkeys tamed in the Middle East; elephants tamed in India
4,500 years ago	Peanuts in the Andes
3,500 years ago	World population is 5–10 million people

CLASSICAL CIVILIZATIONS

3,500 B.C.E.	Sumerian civilization in Mesopotamia: the wheel and cuneiform writing
3,300 B.C.E.	Chronic warfare
3,000 B.C.E.	Civilization of the Nile in Egypt; advances in technology
2,800 B.C.E.	Indus Valley civilization on the Indus River
2,100 B.C.E.	Minoan civilizations on Crete
2,000 B.C.E.	Megalithic structures in Europe
1,700 B.C.E.	Earliest origins of the alphabet in Palestinian region; Aryan-Vedic peoples with Sanskrit language enter India
1,525 B.C.E.	Shang Dynasty in North China
1,200 B.C.E.	Greek settlements. Exodus of Israel from Egypt, monotheism
1,100 B.C.E.	Olmec civilization in Meso-America
700 B.C.E.	Homer
628 B.C.E.	Zoroaster
600 B.C.E.	Beginning of Greek philosophy
560 B.C.E.	Confucius in China; Buddha in India
550 B.C.E.	Persian empire
509 B.C.E.	Founding of Roman Republic
450 B.C.E.	Socrates, Plato, Aristotle
327 B.C.E.	Alexander's invasion of the Indus Valley
260 B.C.E.	India united under Asoka
221 B.C.E.	China united in the empire of Ch'in Shih-huang-ti
150 B.C.E.	Chang Ch'ien establishes route to Bactria
31 B.C.E.	Roman empire under Augustus Caesar
4 B.C.E.	Jesus
64 C.E.	Buddhism in China
100 C.E.	World population is 300 million
300 C.E.	Classical Mayan civilization

410 C.E.	Fall of Rome
650 C.E.	Muslim empire
732 C.E.	Muslim advance in Europe stopped at Poitiers in France
800 C.E.	Carolingian renaissance in Europe; beginning of medieval civilization
900 C.E.	Toltec empire
925 C.E.	Arabic numerals
1000 C.E.	Islamic science
1095 C.E.	Crusades
1115 C.E.	Compass invented
1200 C.E.	Inca empire
1211 C.E.	Beginning of the Mongolian empire under Ghenghis Khan
1271 C.E.	Marco Polo begins travels
1320 C.E.	Aztec empire
1325 C.E.	Ibn Battuta begins journeys
1347 C.E.	Black Death, European population declines
1433 C.E.	Cheng Ho completes voyages to the India Ocean to the Persian Gulf
1453 C.E.	Constantinople falls to Turks
1492 C.E.	Columbus sails to America
1500 C.E.	World population is 400–500 million
1607 C.E.	English settlement of North America begins at Jamestown

THE RISE OF NATIONS

1600 C.E.	British East India Company chartered
1623 C.E.	Japanese policy of isolation
1757 C.E.	British control over India
1763 C.E.	European powers divide the colonial world
1776 C.E.	American Revolution
1789 C.E.	French Revolution
1841 C.E.	The Opium War settled with China establishing five trading ports
1854 C.E.	Perry forces Japan open to western trade
1884 C.E.	European powers divide Africa into European colonies
1914 C.E.	World War I
1917 C.E.	Communism takes control in Russia

1919 C.E.	League of Nations
1939 C.E.	World War II
1945 C.E.	First atomic bomb exploded over Hiroshima; United Nations charter
1982 C.E.	World Charter for Nature
1991 C.E.	Dissolution of Soviet Union
1992 C.E.	UN Conference on Environment and Development

THE MODERN REVELATION

1543 C.E.	Copernican revolution
1609 C.E.	Johannes Kepler discovers the elliptical movement of the planets around the sun
1609 C.E.	Galileo establishes empirical mode of observation by effectively using precise measurements in his observations of natural phenomena
1620 C.E.	Francis Bacon promotes a pragmatic orientation of modern science
1637 C.E.	René Descartes establishes mathematic mode of dealing with the natural world and divides the physical world and mind into two entirely different realms
1687 C.E.	Isaac Newton explains the modern view of the universe
1749 C.E.	Georges-Louis Buffon rethinks the Age of the Earth
1750 C.E.	Carolus Linnaeus provides the modern system of taxonomic classification of life
1755 C.E.	Immanuel Kant proposes a theory of the formation of celestial bodies and the solar system
1795 C.E.	James Hutton discovers that the geological formation of the Earth and of life can be traced back in time
1809 C.E.	Jean-Baptiste Lamarck traces the evolutionary sequence from lower forms to higher forms of life
1827 C.E.	Baron Georges Cuvier sets the basis for the classification of animals
1830 C.E.	Sir Charles Lyell describes the structure of the Earth

277

1859 C.E.	Charles Darwin publishes his theory of natural selection and alters our understanding of the development of life
1905 C.E.	Albert Einstein alters our basic understanding of time, space, motion, matter, and energy
1927 C.E.	Werner Heisenberg changes our perception of knowledge at the atomic level
1929 C.E.	Edwin Hubble provides evidence that we live in an expanding universe
1950 C.E.	Hans Albrecht Bethe describes how stars evolve
1962 C.E.	Rachel Carson exposes the effects of modern pesticides on the natural world
1965 C.E.	Robert Wilson and Arno Penzias find evidence of the origin of the universe

Glossary

algorithm. A precise formulation of a method for doing something. In computers, an algorithm is usually a collection of procedural steps or instructions organized and designed so that computer processing results in the solution of a specific problem.

amino acids. Any of the twenty-five organic molecules that link together into polypeptide chains to form proteins that are necessary for all life.

Argos. The first multicellular animal, a community of formerly autonomous cells now governed by a unifying cybernetic mind. In Greek mythology, Argos was a being with eyes all over his body.

Note: With respect to our familiarity with them, the main actors in the story of the universe fall into three groups. First are those that are well known to us, for instance, carbon, Milky Way galaxy, Sun, Earth, fish, dinosaurs, mammals, paleolithic humans, industrial societies. The second set of actors are those that, though not well known, can be easily identified, such as the primeval Flaring Forth, dwarf stars, prokaryotic cells, eukaryotic cells, Homo habilis, the Neolithic villager. The third set are neither well known nor easily identified, for the simple reason that they have only recently emerged into human awareness. For the most part they do not even have names in the scientific literature. Concerning this third group, our way of proceeding has been to bring forward some ancient names, names that were invented for entirely different purposes, but names that already carry some of the feelings and meanings appropriate for these newly discovered entities. For convenience we list them here: Tiamat, Aries, Prospero, Promethio, Viking, Engla, Vikengla, Kronos, Sappho, Tristan, Iseult, Capaneus, Argos.

Aries. The first living being to arise within the primordial waters of Earth some four billion years ago. The first prokaryotic cell. The name within Egyptian cosmology referred to the thunderbolt that arises out of the primordial waters—the creative spirit at the moment of actualization. (See Note: Argos.)

biome. Any of several major life zones of interrelated plants and animals determined by the climate such as polar ice caps, tundra, taiga, temperate forests, savannah, desert, tropics.

black hole. The final stage of a large star that has collapsed so tightly its gravitational attraction holds even its own photons from escaping.

Canpaneus. The first living being to climb out of the seas and live upon the land. The plant that invented the structure necessary for standing up in a gravitational field. In Greek mythology, Canpaneus is a fearless warrior ready to overcome all limitations, even those set down by the heavens. (See Note: Argos.)

cellular cytoplasm. The protoplasm of a cell, outside the nucleus.

chromosomal DNA. Structural carrier of hereditary characteristics, found in the cell nucleus. Chromosomes are made of protein and nucleic acid.

cold dark matter. Nonluminous matter comprising the bulk of the universe, known solely through its gravitational effects.

conscious self-awareness. Human faculty of understanding characterized by its sense of wonder and celebration as well as by its ability to refashion and use its exterior environment as instruments in achieving its own ends.

curvature in the space-time manifold. According to the general theory of relativity, space-time becomes curved in the vicinity of matter; the greater the concentration of matter, the greater the curvature.

density arms. Shock waves creating the arms in a spiral galaxy.

developmental universe. A universe emerging into being over fifteen billion years of evolutionary process.

DNA, deoxyribonucleic acid. A molecule found in the nuclei of cells. It is the principal constituent of chromosomes, the structures that transmit hereditary characteristics.

Earth Community. The interacting complexity of all of Earth's components, entities, and processes, including the atmosphere, hydrosphere, geosphere, biosphere, and mindsphere.

Ecozoic. The emerging period of life following the Cenozoic, and characterized, at a basic level, by its mutually enhancing human-Earth relations. The word derives from the scientific tradition that divides the Phanerozoic eon into the Paleozoic, Mesozoic, and Cenozoic eras.

electromagnetic. The interaction responsible for the long-range force of repulsion of like and attraction of unlike electric charges.

elementary particles. Bits of matter assumed to be the most basic constituents of the universe. There are four classes of particles distinguished according to their behavior with the fundamental interactions of the universe: the gravitational, electromagnetic, strong nuclear, and weak nuclear. The gravitational is experienced by all particles. The electromagnetic is experienced only by charged particles, although it is transmitted by the photon, which has no charge. The weak and strong nuclear operate at the atomic level.

Of the four classes of particles, the smallest is that of the massless bosons, which include the photon, eight types of gluons, and the hypothetical graviton. The lepton class includes twelve particles: the electron, the positron, the positive and negative muons, the tauon and its antiparticle, and the neutrino or antineutrino associated with each of these particles. The boson and the leptons are not strongly interacting. Members of the meson class are more massive than the leptons. The mesons are the "glue" that holds nuclei together. By far the largest class of particles is the baryon class, the lightest member of which are the proton and neutron; the heavier members are the hyperons. Baryons and mesons, both strongly interacting, are sometimes considered together as hadrons. Hadrons are built up of other, still more fundamental particles called quarks.

Engla. The large prokaryotic cell that suffered the attacks of the Viking cells. (See Viking and Note: Argos.)

entropy. A measure of disorder or randomness of a physical system; a measure of the incapacity of a system to undergo spontaneous change.

enzyme. Any of various proteins, formed in plant and animal cells, that act as organic catalysts in initiating or speeding up specific chemical reactions.

epigenetic. The way the structures in the universe arise in one time and place, evolve in interaction with the universe, achieve a stable if nonequilibrium process, and then decay and disintegrate.

episome. A small genetic element or unit of DNA that is not essential to the life of the cell: it can be lost or transferred and it can replicate independently.

eukaryote. A cell with a membrane-bound nucleus, membrane-bound organelles, and chromosomes in which DNA is combined with special proteins.

gamete. The haploid reproductive cell whose nucleus fuses with that of another gamete of an opposite sex (fertilization); the resulting cell (zygote) may develop into a new diploid individual, or, in some species, may undergo meiosis to form haploid somatic cells.

gene. The functional unit of heredity; usually a sequence of nucleotides in a DNA molecule that codes for a polypeptide.

genome. One complete (haploid) set of chromosomes of an organism with their associated genes.

glucose. White crystalline sugar found in fruits, honey, and blood. Glucose is the major source of energy in animal metabolism.

gravitational. The interaction that manifests itself as a long-range force of attraction between all elementary particles.

Homo erectus. Human species dating from 1.5 million years ago to 300,000 years ago.

Homo habilis. First human species, appearing around 2.6 million years ago in Africa.

Homo sapiens. The only surviving species of humans, dating back some 300,000 years ago.

ionized. An atom or molecule that has lost or gained one or more electrons, thus becoming electrically charged.

Iseult. The sexual gamete cell created by Sappho and sent forth into the surrounding waters. The first haploid ovum. In Celtic mythology, Iseult is the Irish princess who falls hopelessly in love with the knight Tristan (Suggested by P. Cousineau). (See Note: Argos.)

isotopes. Atom of an element that differs from other atoms of the same element in the number of neutrons in the atomic nucleus; isotopes thus differ in atomic weight. Some isotopes are unstable and emit radioactivity.

Kronos. The first creature to thrive by swallowing whole some of its living neighbors. The eukaryotic cell that discovered heterotrophy. The name in the ancient Greek tradition refers to a being who swallowed his own children alive. (See Note: Argos.)

macrocosm. Large-scale, overarching in the space-time continuum; the universe.

meiotic sex. Sexual reproduction generally requiring two parents that come together to form a new genetic identity (the offspring). Sexual reproduction is characterized by two events: the coming together of the sex cells, or gametes (fertilization) and meiosis. Meiosis involves a special kind of division of a cell's nucleus in which the genetic material (chromosomes) divides in half (twice) into four cells. In humans these cells are known as the ovum and sperm (the gametes or sex cells). Segregation, crossing over, and reassortment of the genes occur. As a result, each cell produced by meiosis has a unique assortment of genetic information.

mesocosm. Intermediate, middle; that layer of the universe that is a combination of the microcosm and macrocosm.

metabolic pathways. The organization of living systems into the building up of organic molecules from simpler ones, and the breaking down of complex substances, often accompanied by the release of energy.

microcosm. Miniature, small-scale; referring to organisms, cells, molecules, atoms, subatomic particles.

nonequilibrium state. A system with free energy, a system capable of spontaneous change.

nucleic acids. Organic substance, found in all living cells, in which the hereditary information is stored and from which it can be transferred. The two types, DNA and RNA, consist of long chains that generally occur in combination with proteins.

nucleotides. The basic building blocks of nucleic acid.

nuclides. A specific type of atom that is characterized by its nuclear properties, such as the number of neutrons and protons and the energy state of its nucleus.

ontogeny. The life cycle of a single organism; biological development of the individual.

phenotype. The manifest characteristics of an organism collectively, including anatomical and psychological traits, that result from both its heredity and its environment.

phyla. A taxonomic grouping of related, similar classes; a high-level category beneath kingdom and above class.

plasma. A fully ionized gas containing approximately equal numbers of positive and negative ions, such as occur in the interior of star and in interstellar gas.

physics, quantum and classical. The branch of science traditionally defined as the study of matter, energy, and the relation between them. Modern physical theory holds that energy and some other physical properties often exist in tiny, discrete amounts, called quanta. Quantum theory and the theory of relativity together form the theoretical basis of modern physics. The first contribution to quantum theory was the explanation of the blackbody radiation in 1900 by Max Planck. In 1905 Albert Einstein, in order to explain the photoelectric effect, proposed that radiation itself is also quantized and consists of light quanta, or photons, that behave like particles. Niels Bohr used the quantum theory in 1913 to explain both atomic structure and atomic spectra. Classical physics includes the traditional branches that were recognized and fairly well developed before the twentieth century: mechanics, the study of motion and the forces that cause it; acoustics, the study of sound; optics, the study of light; thermodynamics, the study of the relationships between heat and other forms of energy; and electricity and magnetism. Most of classical physics is concerned with matter and energy on the normal scale of observation. By contrast, much of modern physics is concerned with the behavior of matter and energy under extreme conditions: low-temperature physics, or on the very small scale, nuclear physics.

plasmid. A small, DNA-containing, self-reproducing cytoplasmic element that exists outside the chromosome, as in some bacteria: because it can alter a hereditary characteristic when introduced into another bacterium, it is used in recombinant DNA technology.

primogenial. The earliest ancestor of a family, race, etc.

prokaryotes. Single-cell organisms without nuclei (bacteria) that were the first life forms of Earth. They predominated for approximately 2 billion years and still compose a large segment of the life community.

Promethio. The living being who was first capable of taking the energy from the sun and thriving on it. The prokaryotic cell that invented the process of photosynthesis. The name derives from Prometheus of Greek cosmology, who procured fire from the heavens for the benefit of his companions on Earth. (See Note: Argos.)

Prospero. The first living being capable of dealing creatively with oxygen. In Shakespeare's *The Tempest,* Prospero transforms his enemies and their devastation into a serene and creative renewal. (See Note: Argos.)

quantum. See physics.

quantum mechanics. The application of quantum theory to the motions of material particles, developed during the 1920s.

replicon. A specific sequence of nucleic acid that replicates as a unit when activated.

Sappho. The first eukaryotic cell to engage in meiotic sex—sexual union involving two complementary sets of genetic information. Named after the famous writer born c. 600 B.C.E., whose beautiful love poetry earned her the title of the Tenth Muse. (See Note: Argos.)

second law of thermodynamics. The law that states that during any process the entropy of a system and its surroundings never decreases.

self-organizing system. Any system that regulates the evolution of its own space-time structures, balancing itself in the midst of generative and degenerative processes.

stellar nucleosynthesis. The process by which, as the core of a star heats up, thermonuclear reactions are ignited and hydrogen is converted into helium. In a similar fashion, all the elements from helium through iron are created.

strong nuclear. Short-range interaction that overshadows all other forces between protons and neutrons.

symbiotic relationship. The intimate living together of two kinds of organisms, especially if such association is of mutual advantage.

theory of general relativity. Einstein's theory equating gravitational dynamics to the motions caused by the curvature of space-time.

Tiamat. The star whose supernova explosion some five billion years ago gave birth to the elements that would form Sun, Earth, Mars, Jupiter, and the other planets. The name is taken from Middle Eastern cosmology, which imagined the world as made from a divine being, Tiamat. Tiamat's body was dismembered; half of Tiamat became Heaven, and half of Tiamat became Earth. (See Note: Argos.)

Tristan. The first sperm cell. (See Iseult and Note: Argos.)

Vikengla. The first eukaryotic cell, the ancestor of all forms of advanced life. This complex cell grew out of the symbiotic relationships worked out between the Viking and the Engla lines. (See Note: Argos.)

Viking. A parasitic form of the Prospero line who learned to survive by burrowing into other living cells and consuming their insides. In the Middle Ages of Europe, the Vikings attacked England (called Engla at that time), eventually penetrating its frontier and settling permanently within the island (Suggested by P. Cousineau). (See Note: Argos.)

weak nuclear. Short-range interaction such as radioactive decay whose characteristic strength is for low-energy phenomena a thousand times weaker than electromagnetic.

283

Bibliography

1. Primordial Flaring Forth

Alfven, Hannes. *Worlds-Antiworlds; Antimatter in Cosmology.* San Francisco: Freeman, 1966. A cosmological model that differs from the Big Bang cosmology assumed in our account.

Borner, Gerhard. *The Early Universe: Facts and Fiction.* New York: Springer-Verlag, 1988. A technical treatise on the overlap between elementary particle physics and mathematical cosmology.

Chaisson, Eric. *The Life Era: Cosmic Selection and Conscious Evolution.* Illustrated by Lola Judith Chaisson. 1st ed. New York: Atlantic Monthly Press, 1987. An astrophysicist's brilliant popular account of the three macrotransitions of the universe: energy to matter; matter to life; life to mind.

Cornell, James, ed. *Bubbles, Voids, and Bumps in Time: The New Cosmology.* New York: Cambridge University Press, 1989. Readable summaries of contemporary cosmological research by leaders in the field.

Davies, Paul. *Superforce: The Search for a Grand Unified Theory of Nature.* New York: Simon and Schuster, 1984.

Herbert, Nick. *Quantum Reality: Beyond the New Physics.* Garden City, NY: Anchor Press/Doubleday, 1985. The best overview of the various philosophical interpretations of quantum physics.

Jastrow, Robert. *Until the Sun Dies.* 1st ed. New York: Norton, 1977. One of the first popular accounts of the story of the universe from the Big Bang to today.

Kafatos, M., and Nadeau, R. *The Conscious Universe: Part and Whole in Modern Physical Theory.* New York: Springer-Verlag, 1990. A discussion of Bell's theorem and the implications of nonlocality.

Lederman, Leon M., and Schramm, David N. *From Quarks to the Cosmos: Tools of Discovery.* New York: Scientific American Library (distributed by Freeman), 1989. Two of America's foremost scientists team up.

Lemaitre, Georges. *The Primeval Atom, an Essay on Cosmogony.* Preface by Ferdinand Gonseth, foreword to the English edition by Henry Norris Russell. Translated by Betty H. and Serge A. Korff. New York: Van Nostrand, 1950. Another alternative cosmological model.

Leslie, John, ed. *Physical Cosmology and Philosophy.* New York: Macmillan, 1990. This book asks scientists and philosophers the controversial questions such as "Was there a Big Bang?" "Was life inevitable in our universe?"

Pagels, Heinz R. *The Cosmic Code: Quantum Physics as the Language of Nature.* New York: Bantam Books, 1983. A popular introduction to the basics of modern physics.

Peat, F. David. *Superstrings and the Search for the Theory of Everything.* Chicago: Contemporary Books, 1988. The contemporary quest for a theory that unifies the four fundamental interactions of the physical universe.

Peat, F. David. *Einstein's Moon: Bell's Theorem and the Curious Quest for Quantum Reality.* Chicago: Contemporary Books, 1990.

Reeves, Hubert. *Atoms of Silence: An Exploration of Cosmic Evolution.* Translated by Ruth A. Lewis and John S. Lewis. Cambridge, MA: Massachusetts Institute of Technology, 1984.

Serway, Raymond A., Moses, Clement J., and Moyer, Curt A. *Modern Physics,* Philadelphia: Saunders College Publishing, 1989. A technical introduction to the basics of modern physics.

Silk, Joseph. *The Big Bang: The Creation and Evolution of the Universe.* Foreword by Dennis Sciama. San Francisco: Freeman, 1980. A comprehensive treatment of all the major topics in cosmology today, written by a researcher in the field, suitable for the first-year college student.

Tryon, Edward. "Is the Universe a Vacuum Fluctuation?" *Nature* 246: 396–397. First reflections on the birth of the universe from the perspective of quantum physics and Heisenberg's principle.

Weinberg, Steven. *The First Three Minutes: A Modern View of the Origin of the Universe.* Updated ed. New York: Basic Books, 1988. Still the best popular treatment of the origin of the universe, by a researcher in the field.

2. Galaxies

Baade, Walter. *Evolution of Stars and Galaxies.* Edited by Cecilia Payne-Gaposchkin. Cambridge, MA: Harvard University Press, 1963.

Bok, Bart Jan, and Bok, Priscilla F. *The Milky Way.* 5th ed. Cambridge, MA: Harvard University Press, 1981.

Campbell, Jeremy. *Grammatical Man: Information, Entropy, Language, and Life.* New York: Simon and Schuster, 1982. Popular introduction to the theories of entropy and information.

Dyson, Freeman. "Energy in the Universe." *Scientific American,* September 1971. Dyson on the density waves and shock waves igniting star birth.

Ferris, Timothy. *Galaxies.* Photographs selected by Timothy Ferris; illustrations by Sarah Landry. New York: Harrison House (distributed by Crown), 1987. Beautiful photographs of every kind of galaxy.

Heisenberg, Werner. *Philosophical Problems of Quantum Physics.* Translated by F. C. Hayes. Woodbridge, CT: Ox Bow Press, 1979.

———. *Physics and Philosophy: The Revolution in Modern Science.* New York: Harper & Row, 1962.

Kron, Richard G., ed. *Evolution of the Universe of Galaxies.* Presented at the Edwin Hubble Centennial Symposium, University of California at Berkeley, 1989, and the Astronomical Society of the Pacific, San Francisco, 1990. Recent technical symposium on the origin and structure of galaxies.

Misner, Charles W., Thorne, Kip. S., and Wheeler, John A. *Gravitation.* San Francisco: Freeman, 1973. Advanced treatise on Einstein's theory of gravitation.

Mitton, Simon. *Exploring the Galaxies.* New York: Scribner's, 1976.

Morrison, Philip, and Morrison, Phylis. *Powers of Ten: A Book About the Relative Size of Things in the Universe and the Effect of Adding Another Zero.* Redding, CT: Scientific American Library (distributed by Freeman, San Francisco), 1982 (also in video). A visual journey from the quark through the mesocosm into the galaxies and large-scale structures of space-time.

Shapley, Harlow. *Galaxies.* 3rd ed. Revised by Paul W. Hodge. Cambridge, MA: Harvard University Press, 1972. Popular introduction by one of the pioneers in twentieth-century astronomy.

Whitehead, Alfred North. *Science and the Modern World.* Cheap ed. Cambridge, England: Cambridge University Press, 1953. Critique of modern science's overemphasis on reductionism and mechanism.

3. Supernovas

Barrow, John D., and Silk, Joseph. *The Left Hand of Creation: The Origin and Evolution of the Expanding Universe.* Boston: Unwin Paperbacks, 1983. A wide-ranging collection of original reflections on the universe.

Clayton, P. *Principles of Stellar Evolution and Nucleosynthesis.* Chicago: University of Chicago Press, 1983. Advanced treatise.

Davies, P. C. W. *The Accidental Universe*. New York: Cambridge University Press, 1982. A mathematical introduction to cosmic coincidences, including the subtle relationships enabling supernovas to exist, suitable for college science students.

Goldsmith, Donald. *Supernova! The Exploding Star of 1987*. 1st ed. New York: St. Martin's Press, 1989.

Kippenhahn, Rudolf. *100 Billion Suns: The Birth, Life, and Death of the Stars*. Translated by Jean Steinberg. New York: Basic Books, 1983. Popular account by a researcher in the field.

Marschall, Laurence A. *The Supernova Story*. New York: Plenum Press, 1988. Readable introduction to supernovae in general and Supernova 1987A in particular.

Murdin, Paul. *End in Fire: The Supernova in the Large Magellanic Cloud*. New York: Cambridge University Press, 1990.

Murdin, Paul, and Murdin, Lesley. *Supernovae*. Rev. ed. New York: Cambridge University Press, 1985.

4. Sun

Bonner, John Tyler. *Morphogenesis: An Essay on Development*. New York, Atheneum, 1963. Accessible introduction.

Dermott, S. F., ed. *The Origin of the Solar System*. Chichester, NY: Wiley, 1978.

Hatsopoulos, George N., and Keenan, Joseph H. *Principles of General Thermodynamics*. Malabar, FL: Krieger, 1965. Advanced treatise covering the second law of thermodynamics, nonequilibrium systems, and information theory.

Hawking, Stephen W. *A Brief History of Time: From the Big Bang to Black Holes*. Introduction by Carl Sagan; illustrations by Ron Miller. London: Bantam, 1988. Brilliant sections on black holes by the world's most famous cosmologist.

Jantsch, Erich. *The Self-Organizing Universe: Scientific and Human Implications of the Emerging Paradigm of Evolution*. 1st ed. New York: Pergamon Press, 1980. The best comprehensive treatment of the change in cosmological orientation arising out of Prigogine's work in autopoietic systems.

Jantsch, Erich, ed. *The Evolutionary Vision: Toward a Unifying Paradigm of Physical, Biological, and Sociocultural Evolution*. Boulder, CO: Westview Press, 1981.

Jantsch, Erich, and Waddington, Conrad, eds. *Evolution and Consciousness: Human Systems in Transition*. Reading, MA: Addison-Wesley, 1976.

Laszlo, Ervin. *Evolution: The Grand Synthesis*. Foreword by Jonas Salk. Boston: New Science Library, 1987. From the systems perspective.

Miller, James Grier. *Living Systems*. New York: McGraw-Hill, 1978. The standard treatise on systems science.

Nitecki, Matthew H., ed. *Evolutionary Progress*. Chicago: University of Chicago Press, 1988. Distinguished scientists grappling with the question of progress in the universe.

Prigogine, Ilya. *From Being to Becoming: Time and Complexity in the Physical Sciences*. San Francisco: Freeman, 1980.

Prigogine, Ilya, and Nicolis, Gregoire. *Self-Organization in Non-Equilibrium Systems*. New York: Wiley-Interscience, 1977. Seminal works in the dynamics of self-organization.

Thompson, D'Arcy Wentworth. *On Growth and Form*. Abridged ed. Edited by John Tyler Bonner. Cambridge, England: Cambridge University Press, 1969. Highly original foundational work on the emergence and development of form.

Wood, John A. *The Solar System*. Englewood Cliffs, NJ: Prentice-Hall, 1979.

5. Living Earth

Bateson, Gregory. *Mind and Nature: A Necessary Unity*. New York: Bantam Books, 1988. A cybernetic account that treats biological evolution and ecosystemic interactions as mental processes.

Calder, Nigel. *The Restless Earth: A Report on the New Geology*. New York: Penguin Books, 1972.

————. *The Life Game*. New York: Viking, 1973. Evolution and the new biology.

————. *Timescale: An Atlas of the Fourth Dimension*. New York: Viking, 1983. In addition to summarizing valuable information on early Earth, provides an excellent and concise account of the entire evolutionary story. We have for the most part based our dating on the chronology of *Timescale*.

Cotterill, Rodney. *The Cambridge Guide to the Material World*. New York: Cambridge University Press, 1985. Beautiful photographs to accompany text on Earth's material structures.

Eigen, Mangred. "Self-Organization of Matter and the Evolution of Biological Macromolecules," *Naturwissenschaften* 58: 465–523. Application of concepts of autopoiesis to the evolutionary process leading to life.

Jastrow, Robert. *Red Giants and White Dwarfs: Man's Descent from the Stars*. Rev. ed. New York: Harper & Row, 1971. Popular account of the origin and development of solar system, with chapters on Mars and Jupiter.

Lovelock, James E. *Gaia: A New Look at Life on Earth*. New York: Oxford University Press, 1979. The first scientific formulation of the hypothesis that Earth is a living system.

Margulis, Lynn, and Sagan, Dorian. *Microcosmos: Four Billion Years of Evolution from Our Microbial Ancestors*. Foreword by Lewis Thomas. 1st Touchstone ed. New York: Simon and Schuster, 1986. Brilliant account of the age of bacteria, including all major creative developments.

Miller, Stanley L., and Orgel, Leslie E. *The Origins of Life on the Earth*. Englewood Cliffs, NJ: Prentice-Hall, 1974. Summary statement by two leading researchers.

Oparin, Aleksandr Ivanovich. *The Origin of Life*. Translated with annotations by Sergius Morgulis. 2d ed. New York: Dover Publications, 1953. The first modern scientific account of the origin of life.

Smith, David G., ed. *The Cambridge Encyclopedia of Earth Sciences*. New York: Crown, 1981. Comprehensive treatment of the physical structures of Earth.

Stolz, John, ed. *Structure of Phototrophic Prokaryotes*. Boca Raton, FL: CRC Press, 1991.

Woese, C. R., and Fox, G. E. *Proceedings of the National Academy of Sciences* 74 (1977): 5088. The fundamental division of living things traced to molecular evolution.

6. Eukaryotes

Curtis, Helena. *Biology*. 3rd ed. New York: Worth Publishers, 1979. One of the best college textbooks covering all major topics in biology.

Dott, Robert H., and Batten, Lyman. *Evolution of the Earth*. 4th ed. New York: McGraw-Hill, 1988. The standard treatise.

Eigen, M., and Schuster, P. *The Hypercycle: A Principle of Natural Self-Organization*. New York: Springer-Verlag, 1979. The process used by the early Earth for building up complexity, including multicellularity.

Laszlo, Ervin. *The Systems View of the World: The Natural Philosophy of the New Developments in the Sciences*. New York: G. Braziller, 1972.

Lovelock, J. E. *The Ages of Gaia: A Biography of Our Living Earth*. 1st ed. New York: Norton, 1988. A new biography of Earth, told from the perspective that Earth is a living system.

Margulis, Lynn. *Origin of Eukaryotic Cells: Evidence and Research Implications for a Theory of the Origin and Evolution of Microbial, Plant, and Animal Cells on the Precambrian Earth*. New Haven, CT: Yale University Press, 1970.

————. *Symbiosis in Cell Evolution: Life and Its Environment on the Early Earth*. San Francisco: Freeman, 1981. The new understanding of symbiosis in cellular evolution by a leading researcher.

Margulis, Lynn, ed. *Handbook of Protoctista: Structure, Cultivation, Habitats, and Life Histories of the Eukaryotic Microorganisms and Their Descendents Exclusive of Animals, Plants, and Fungi*. Boston: Jones and Bartlett, 1990.

Margulis, Lynn, and Sagan, Dorian. *Origins of Sex: Three Billion Years of Genetic Recombination*. New Haven, CT: Yale University Press, 1986.

Margulis, Lynn, and Schwartz, Karlene V. *Five Kingdoms: An Illustrated Guide to the Phyla of Life on Earth*. 2nd ed. New York: Freeman, 1988. In addition to plants and animals, the authors give due attention to the three other kingdoms: fungi, protists, and bacteria.

Monod, Jacques. *Chance and Necessity*. New York: Vintage Books, 1971, p. 126.

Raup, David M., and Stanley, Steven M. *Principles of Paleontology*. 2nd ed. San Francisco: Freeman, 1978.

Sahtouris, Elisabet. *Gaia: The Human Journey from Chaos to Cosmos*. New York: Pocket Books, 1989. A philosopher looks at the Gaia hypothesis and its implications.

7. Plants and Animals

Birch, Charles, and Cobb, John B. *The Liberation of Life: From the Cell to the Community*. New York: Cambridge University Press, 1981. A biologist and a philosopher's account in which the subjectivity of an organism receives full attention.

Calvin, William H. *The River That Flows Uphill: A Journey from the Big Bang to the Big Brain*. New York: Macmillan, 1986. A fascinating popular account of the evolutionary story, told by a neuroscientist as he rafts down the Grand Canyon.

Cobb, John, and Griffin, David, eds. *Mind in Nature: Essays on the Interface of Science and Philosophy*. Contributions by Charles Birch et al. Washington, DC: University Press of America, 1977.

Darwin, Charles. *The Origin of Species by Means of Natural Selection; or, The Preservation of Favoured Races in the Struggle for Life*. New foreword by George Gaylord Simpson. New York: Collier Books, 1962. After a century of mining, this work still yields treasures.

Dobzhansky, Theodosius Grigorievich. *Genetics and the Origin of Species*. 3rd ed., rev. New York: Columbia University Press, 1964. By one of the founders of the synthetic theory.

Ehrlich, Paul R., Ehrlich, Anne H., and Hodren, John P. *Ecoscience: Population, Resources, Environment*. 3rd ed. San Francisco: Freeman, 1977. Comprehensive treatise.

Eldredge, Niles. *Macroevolutionary Dynamics: Species, Niches, and Adaptive Peaks*. New York: McGraw-Hill, 1989.

Futuyma, Douglas J. *Evolutionary Biology*. 2nd ed. Sunderland, MA: Sinauer Associates, 1986. With excellent references to the contemporary literature.

Ghiselin, Michael T. *The Triumph of the Darwinian Method*. Chicago: University of Chicago Press, 1969. An explication of Darwin's life work by a prominent biologist and philosopher of biology.

Gould, Stephen Jay. *Wonderful Life: The Burgess Shale and the Nature of History*. 1st ed. New York: Norton, 1989. A leading paleontolgist recounts the discovery of some of the most significant fossils of the twentieth century.

Hull, David L. *The Metaphysics of Evolution*. Albany: State University of New York Press, 1989. Summary statement by one of the foremost philosophers of biology.

Mayr, Ernst. *The Growth of Biological Thought: Diversity, Evolution, and Inheritance*. Cambridge, MA: Belknap Press, 1982. Comprehensive treatise by a master biologist.

Monod, Jacques. *Chance and Necessity: An Essay on the Natural Philosophy of Modern Biology*. Translated by Austryn Wainhouse. New York: Vintage Books, 1971. Philosophical reflections by a major figure in twentieth-century biology.

Ruse, Michael. *Philosophy of Biology Today*. Albany: State University of New York Press, 1988. While physics dominated the philosophy of science during the nineteenth and early twentieth centuries, biology now makes its claim as the central science of our time. This volume is a guidebook to some of the best thinking in philosophy of biology.

Simpson, George Gaylord. *The Meaning of Evolution: A Study of the History of Life and of Its Significance for Man*. Rev. ed. New Haven, CT: Yale University Press, 1967.

Stanley, Steven M. *Earth and Life Through Time*. 2nd ed. New York: Freeman, 1989. The first comprehensive account of the history of Earth that combines the ecological and evolutionary perspectives.

Stanley, Steven M. *Extinction*. New York: Scientific American Library (distributed by Freeman), 1987. The contours of the major extinctions over the last billion years.

Wilson, Edward Osborne. *Sociobiology: The New Synthesis*. Cambridge, MA: Belknap Press of Harvard University Press, 1975.

8. Human Emergence

Brown, Michael H. *The Search for Eve*. New York: Harper & Row, 1990. Popular summary of the present data.

Campbell, Bernard G. *Humankind Emerging*. 5th ed. Boston: Scott, Foresman, 1988. A scholarly presentaion of the evolutionary sequence leading to Homo sapiens.

Eliade, Mircea. *Shamanism: Archaic Techniques of Ecstasy*. Translated by Willard R. Trask. Rev. and enl. ed. Princeton, NJ: Princeton University Press, 1974. The classic work on the shamanic rapport with the spirit powers of the universe.

Grim, John A. *The Shaman: Patterns of Siberian and Ojibway Healing*. Norman, OK: University of Oklahoma Press, 1984. Explains the functional role of cosmology in indigenous tribal rituals.

Hawkes, Jacquetta. *Prehistory*. History of Mankind Series: Cultural and Scientific Development, vol. 1, part 1. New York: Harper & Row, 1963. A brilliant overview of the geological context as well as the physical and cultural phases of early human development.

Johanson, Donald, and Edey, Maitland. *Lucy: The Beginning of Humankind*. New York: Warner Books, 1982. A firsthand report on one of the most remarkable hominid discoveries and her place in the larger pattern of human development.

Pfeiffer, John E. *The Creative Explosion: An Inquiry Into the Origins of Art and Religion*. 1st ed. New York: Harper & Row, 1982. A detailed study of the spiritual-religious experience of early humans.

Stanley, Steven M. *The New Evolutionary Timetable*. New York: Basic Books, 1981.

White, Randall. *Dark Caves, Bright Visions: Life in Ice Age Europe*. New York: Norton, 1986. Comprehensive and well-presented account of Upper Paleolithic art, with excellent photographs.

9. Neolithic Village

Eisler, Riane. *The Chalice and the Blade: Our History Our Future*. San Francisco: Harper & Row, 1987. The human situation in Neolithic times, with special attention given to the social roles women occupied, and the implications for gender relations in our time.

Eliade, Mircea. *The Myth of the Eternal Return; or Cosmos and History*. Translated by Willard R. Trask. Princeton, NJ: Princeton University Press, 1974. The basic myth of the universe as the context for human self-discovery.

Gimbutas, Marija Alseikaite. *The Civilization of the Goddess: The World of Old Europe*. Edited by Joan Marler. San Francisco: Harper San Francisco, 1991. A remarkable new interpretation of the early Neolithic, the role of women, and the change of social orientation from a feminine emphasis to a masculine emphasis based on archeological research.

Hadingham, Evan. *Early Man and the Cosmos*. Norman: University of Oklahoma Press, 1984. A brief but impressive and very clear survey of early human presence to cosmic powers.

Levy, Gertrude Rachel. *The Gate of Horn: A Study of the Religious Conceptions of the Stone Age and Their Influence upon European Thought*. New York: Book Collectors Society, 1948; London: Faber, 1963. The earliest and still valid study of the religious mystique of the cave art of the late Paleolithic and early Neolithic.

Mellart, James. *The Neolithic of the Near East*. New York: Scribner's, 1975. A reassessment of the Neolithic in Asia Minor based on excavations principally at Çatal Hüyük.

Spretnak, Charlene. *Lost Goddesses of Early Greece: A Collection of Pre-Helenic Myths*. Berkeley, CA: Moon Books, 1978. A small collection of the myths of feminine deities in the pre-Aryan period of early Greece.

10. Classical Civilizations

Ali, Syed Ameer. *The Spirit of Islam*. London: Christophers, 1922. A critical self-analysis and effort at self-identification from within Islam itself, early in the twentieth century.

Carrasco, David. *Quetzalcoatl and the Irony of Empire: Myths and Prophecies in the Aztec Tradition.* Chicago: University of Chicago Press, 1982.

Curtin, Philip D., et al. *African History.* Boston: Little, Brown, 1978. A reliable account of Africa throughout the period dominated by the Eurasian civilizations.

Dawson, Christopher. *Religion and Culture.* New York: Sheed & Ward, 1948. The finest and still among the most impressive of the twentieth-century scholars to insist on the primacy of the religious dynamism in the interpretation of cultural development.

De Bary, William T., Chan, Wing-tsit, and Watson, Barton, eds. *Sources of Chinese Tradition.* New York: Columbia University Press, 1960; De Bary, William T., Tsunoda, Ryusaku, and Keene, Donald, eds. *Sources of Japanese Tradition.* New York: Columbia University Press, 1958; De Bary, William T., et al., eds. *Sources of Indian Tradition.* New York: Columbia University Press, 1958. The finest collection of source materials available in English on these three traditions, with clearly written explanations of the texts chosen.

Driver, Harold E., ed. *The Americas on the Eve of Discovery.* Englewood Cliffs, NJ: Prentice-Hall, 1964. A very readable essay for understanding this historical moment in the context of the indigenous peoples.

Frankfort, Henri. *Before Philosophy, the Intellectual Adventure of Ancient Man: An Essay on Speculative Thought in the Ancient Near East.* Baltimore, MD: Penguin Books, 1966.

Helms, Mary W. *Middle America: A Culture History of Heartland and Frontiers.* Washington, DC: University Press of America, 1975.

Holt, P. M., et al., eds. *The Cambridge History of Islam.* Vol. 1A. Cambridge: Cambridge University Press, 1970.

Khaldoun, Ibn. *The Muqaddimah: An Introduction to History.* 3 vols. 2nd ed. Original Arabic text, 1396. Translated into English by Franz Rosenthal. Princeton, NJ: Princeton University Press, 1967. One of the monumental works of historical interpretation from within Islam.

Las Casas, Bartolome de. *History of the Indes.* Translated and edited by André M. Collard. New York: Harper, 1971. Originally published as *Tyrannies et cruautez des Espagnols, perpetrees e's Indes Occidentales,* 1579. The first and still the most forceful protest against the violence done to the indigenous peoples of the Americas by the Spanish invaders.

Leon-Portilla, Miguel. *Native Meso-American Spirituality.* New York: Paulist Press, 1980.

McNeill, William Hardy. *The Rise of the West: A History of the Human Community.* Drawings by Bela Petheo. Chicago: University of Chicago Press, 1963. The best available single-volume presentation of human history and the dominance of the West in recent centuries.

_____. *Plagues and Peoples.* 1st ed. Garden City, NY: Anchor Press/Doubleday, 1976. A unique and extremely valuable explanation of the influence of disease on human history.

Mumford, Lewis. *The City in History: Its Origins, Its Transformations, and Its Prospects.* 1st ed. New York: Harcourt, Brace and World, 1961. A masterful survey of the structure and functioning of cities from classical to modern times in the West.

Needham, Joseph. *Science and Civilisation in China.* 6 vols. Cambridge, England: Cambridge University Press, 1954–1988. A comprehensive account of science and technology in China in association with Chinese philosophical interpretation of the universe.

Oliver, Roland, and Fage, J. D., eds. *The Cambridge History of Africa.* 8 vols. New York: Cambridge University Press, 1975–1986. A comprehensive source for information on the course of African history from its beginning to modern times.

Radhakrishnan, Sarvepalli, et al., eds. *The Cultural Heritage of India.* 2nd ed. 4 vols. Calcutta: The Ramakrishna Mission, Institute of Culture, 1958–?. A basic study by Indian scholars.

Steward, Julian Haynes, ed. *Handbook of South American Indians.* Smithsonian Institution Bulletin Series, 7 vols. New York, Cooper Square Publishers, 1963–.

Sullivan, Lawrence Eugene. *Icanchu's Drum: An Orientation to Meaning in South American Religions.* New York: Macmillan, 1988. The first comprehensive treatment of the rich indigenous consciousness of the South American continent.

Tedlock, Barbara. *Time and the Highland Maya.* Albuquerque: University of New Mexico, 1982. A look at the sense of time in this most classical of Meso-American civilizations.

Watt, W. M. *Influence of Islam on Medieval Europe.* Edinburgh: Edinburgh University Press, 1972.

Wooley, Sir Leonard. *The Beginnings of Civilization*. History of Mankind Series, vol. 1: Cultural and Scientific Development, part 2. New York: Harper & Row, 1965. A basic source for interpreting the civilizations of the Near East and Egypt.

11. Rise of Nations

Bull, Hedley, and Watson, Adam, eds. *The Expansion of International Society*. Oxford: Clarendon Press, 1984. The tendency of modern societies to articulate themselves as nations and join in the complex of international bonds.

Burckhardt, Jacob. *Force and Freedom: Reflections on History*. New York: Pantheon Books, 1943. A special study by the most impressive eighteenth-century historian of culture. Deals extensively with Europe in the eighteenth and nineteenth centuries C.E.

Freund, Bill. *The Making of Contemporary Africa: The Development of African Society Since 1800*. Bloomington: Indiana University Press, 1984. An overview of modern developments in nineteenth- and twentieth-century Africa.

Hayes, Carlton J. H. *Nationalism: A Religion*. The first American historian to focus his work so clearly on nationalism, both its nature and its consequences.

Kohn, Hans. *The Idea of Nationalism*. New York: Harper & Row, 1962. A brief presentation on the functioning of nationalism.

Lach, Donald. *Asia in the Making of Europe*. 3 vols. Chicago: University of Chicago Press. An invaluable resource for study of the influence of Asia on European civilization from 1500 to 1800 C.E.

Lenski, Gerhard, and Lenski, Jean. *Human Societies: An Introduction to Macrosociology*. 4th ed. New York: McGraw-Hill, 1982.

Pakenham, Thomas. *The Scramble for Africa, 1876–1912*. 1st U.S. ed. New York: Random House, 1991. An account of the sudden European interest in Africa and the competition for occupying the various territories.

Spengler, Oswald. *The Decline of the West*. Authorized translation with notes by Charles Francis Atkinson. New York: Alfred A. Knopf, 1980. The first and most powerfully reasoned presentation of western culture having passed its creative phase and entering into a period of decline.

Sykes, Sir Percy. *A History of Exploration: From the Earliest Times to the Present Day*. 3rd ed. London: Routledge and Kegan Paul, 1949.

Voegelin, Eric. *From Enlightenment to Revolution*. Edited by John H. Hallowell. Durham, NC: Duke University Press, 1975. A masterful presentation of the intellectual and social forces at work in Europe in the seventeenth and eighteenth centuries.

12. The Modern Revelation

Bergson, Henri. *Creative Evolution*. Originally published as *Evolution creatrice* in 1907. Translated by Arthur Mitchell. Westport, CT: Greenwood Press, 1975. A basic work for understanding the emergent universe.

Koyre, Alexandre. *From the Closed World to the Infinite Universe*. Baltimore, MD: Johns Hopkins Press, 1957. The dramatic story of the transition out of the medieval worldview.

Mason, Stephen Finney. *A History of the Sciences*. New rev. ed. New York: Collier Books, 1968. A useful summary of modern scientific development.

Mayr, Ernst. *The Growth of Biological Thought: Diversity, Evolution, and Inheritance*. Cambridge, MA: The Belknap Press of Harvard University Press, 1982. An outstanding presentation of modern biological interpretation.

Merton, Robert K. *Science, Technology and Society in Seventeenth-Century England*. Atlantic Highlands, NJ: Humanities Press, 1978. A thorough sociological survey.

Pais, Abraham. *"Subtle is the Lord—": The Science and the Life of Albert Einstein*. New York: Oxford University Press, 1982. Both the life and the thought of this great figure in twentieth-century scientific development.

Segre, Emilio. *From X-rays to Quarks: Modern Physicists and Their Discoveries*. San Francisco: Freeman, 1980.

Teilhard de Chardin, Pierre. *The Phenomenon of Man*. Originally published as *Le Phenomeon humain* by Editions du Seuil, Paris. Translated by Bernard Wall. New York: Harper & Row, 1959. A narrative of the emergent universe that assumes it had a psychic as well as a physical dimension from the beginning.

Toulmin, Stephen Edelston. *The Return to Cosmology: Postmodern Science and the Theology of Nature*. Berkeley: University of California Press, 1982. A leading philosopher of modern science identifies the significance in our times of the rebirth of cosmology.

Toulmin, Stephen, and Goodfield, June. *The Discovery of Time*. Chicago: University of Chicago Press, 1965.

Whitehead, Alfred North. *Process and Reality: An Essay in Cosmology*. Corrected ed., edited by David Ray Griffin, Donald W. Sherburne. New York: Free Press, 1929. A comprehensive metaphysics taking into account twentieth-century physics.

13. The Ecozoic Era

Bertell, Rosalie. *No Immediate Danger: Prognosis for a Radioactive Earth*. Toronto: Women's Educational Press, 1985.

Berry, Thomas. *The Dream of the Earth*. San Francisco: Sierra Club Books, 1988. Essays toward a new mode of human presence on the Earth that would be mutually enhancing.

Carson, Rachel. *Silent Spring*. Twenty-fifth anniversary ed. Boston: Houghton Mifflin, 1987. The first startling presentation of the chemical poisoning of the land with DDT and its consequences.

Daly, Herman E. *Economics, Ecology, Ethics: Essays Toward a Steady-State Economy*. Edited by Herman E. Daly. San Francisco: Freeman, 1980. A new vision of economic balance with the natural world.

Diamond, Irene and Gloria Feman Orenstein. *Reweaving the World: The Emergence of Ecofeminism*. San Francisco: Sierra Club Books, 1990.

Ellul, Jacques. *The Technological Society*. Translated by John Wilkinson. Introduction by Robert K. Merton. 1st American ed. New York: Knopf, 1964. A powerful critique of mechanistic technologies and their deleterious influences on the human dimension of life.

Gore, Al. *Earth in the Balance: Ecology and the Human Spirit*. New York: Houghton Mifflin, 1992. Possibly the finest analysis by any American political personality since World War II of our disastrous economic and ecological impasse, with hope and guidance for the future.

Fox, Matthew. *Original Blessing*. Santa Fe: Bear and Company, 1986. A basic corrective for the excessive Christian emphasis on original sin and redemption.

Griffin, Susan. *Women and Nature: The Roaring Inside Her*. New York: Harper & Row, 1978. One of the earliest and most basic studies of this subject.

Henderson, Hazel. *Paradigms in Progress: Life Beyond Economics*. Indianapolis: Knowledge Systems Incorporated, 1991. A view of possibilities before us by an economist fully aware of the ecological difficulties of the late twentieth century.

Hyams, Edward. *Soil and Civilization*. New York: State Mutual Books, 1980. A valuable survey of human societies and their effect on the environment. All civilizations from their origins have put considerable stress on the natural systems.

Leopold, Aldo. *A Sand County Almanac*. New York: Oxford University Press, 1949. The classic essay entitled "A Land Ethic" is found in this book.

Meadows, Donella H. et al. *Limits to Growth: A Report to the Club of Rome's Project on the Predicament of Mankind*. 2nd ed. New York: Universe, 1974.

Merchant, Carolyn. *The Death of Nature: Women, Ecology and the Scientific Revolution*. San Francisco: Harper & Row, 1981. A basic source for understanding the sources of our plundering attitude toward the natural world.

Milbrath, Lester W. *Environmentalists: Vanguard for a New Society*. Albany: State University of New York Press, 1984. A sociological study of the role of environmentalists in shaping the social and political destinies of the human community.

Myers, Norman. *Gaia: An Atlas of Planet Management*. With Uma Rath Nath and Melvin Westlake. 1st ed. Garden City, NY: Anchor Press/Doubleday, 1984. A most useful reference work on the condition of the planet in the late twentieth century. Exceptionally well illustrated.

Nash, Roderick. *Wilderness and the American Mind*. 3rd ed. New Haven, CT: Yale University Press, 1982. The basic study of the mystique of the wilderness as this is found in America.

Register, Richard. *Ecocity Berkeley: Building Cities for a Healthy Future*. Berkeley: North Atlantic Books, 1987. A realistic program for renewing our cities within their natural environments through a sequence of transformational stages.

Spretnak, Charlene. *States of Grace: The Recovery of Meaning in the Post-Modern World; Reclaiming the Core Teachings, Practices of the Great Wisdom Traditions*, and *The Well-Being of the Earth Community*. San Francsico: Harper San Francisco, 1991. The modern relevance of ancient traditions.

Worster, Donald. *Nature's Economy: The Roots of Ecology*. Garden City, NY: Anchor Press/Doubleday, 1977. The best source for understanding the historical development of ecological consciousness since the seventeenth century.

Wilson, Edward O. *Biophilia*. Cambridge, MA: Harvard University Press, 1984. A superb biologist deals with the human presence to the planet Earth in all the magnificence of its living forms.

————. ed. *Biodiversity*. Washington, DC: National Academy Press, 1988. A collection of over fifty essays on biodiversity and its role in the integral functioning of the Earth by distinguished scholars.

Acknowledgments

Thomas and I began our conversation in 1982, in the middle of the coldest winter in Chicago's history—Thomas from the South, a child of the Carolinas, a student of cultural history; myself from the North, born in Seattle, a student of the natural sciences. And although he would return to the East to his place in New York and I to the West to live in California, our common probing for a new story of the universe and the role of the human in the story began there under the spacious skies in the middle of our continent.

From the beginning we regarded the articulation of a new cosmology as a task for the species as a whole. What was clear to both of us was that this story is being told by the universe—by the galaxies, by the birds, by the Earth, by the winds, by the stellar explosions, by the fossils, by the rising and falling of the mountain ranges, by the children of every species. Our primary task was to learn how to listen, and to establish rapport with others who are listening. No one person or culture or intellectual discipline by itself had the capacity to hear the full story the universe was telling.

Needed was an institutional enterprise where this probing could be carried out with a sustained discipline and an energetic commitment over a long period of time. Ideal would be a setting in the American system of higher education where scientists, artists, and philosophers, where representatives from the full spectrum of the planet's cultural traditions, could gather together with the common endeavor of listening to and articulating the story of the universe and the role of the human within the universe.

Thanks to the vision and generosity of Laurance S. Rockefeller and his associates George Lamb, Jean and Sidney Lanier, and Elizabeth McCormack, such an institutional setting came into existence—the Center for the Story of the Universe. Without such support and encourgement, this book could easily have remained another interesting but unrealized vision. With the founding of the center, affiliated with the California Institute of Integral Studies, we now have a forum where humans with radically differing capacities for listening to the story can join together in the adventure of seeking a new narrative account of where we have come from, where we are going, and what we are about as a species.

We are grateful for the opportunity to bring many original voices into our explorations at the center: fresh from Africa, Jane Goodall presents the cosmological orientation she learned from the chimpanzees; poet Phil Cousineau shares his research in the powers of the universe as understood in the ancient mythologies; paleontologist Stephen Jay Gould contributes his observations on the fossils from the Burgess Shale and what they teach us about the meaning of life; human ecologist Joe Meeker leads us into the rich world of Dante to articulate what a truly integral cosmology consists of; anthropologist Jim Swan celebrates the Miwok and the Salish peoples and their understanding of the spirit of place; ecofeminist Charlene Spretnak tells the story of the origins of patriarchy in the Kurgan invasions into Old Europe; philosopher of religion Robert McDermott details the evolution of consciousess as understood in the western esoteric traditions; animal behaviorist Robert Fagen comes to the center from Alaska to share with us what he has learned while watching brown bears for most of a decade; Tibetan scholar Robert Thurman brings what he has learned from the indigenous peoples of Tibet concerning the evolution of the life forms; and physicist Minh Duong-van explores with us the effects chaos dynamics have on the unfolding structures of the cosmos.

In our probing for an integral story of the universe we have consulted and argued with, listened to, interviewed, been criticized and corrected and encouraged by a great many thinkers. Often this has involved written responses to various sections of the narrative in our developing manuscript. The passion of so many of these humans for the truths they have sought to impress upon us says a great deal about the deep urgency within the soul for a common story. We remember and thank especially John Giannini, Dave McCloskey, Lavinia Currier, Paul Caringella, Maia Aprahamian, Bea Briggs, Robert Mueller, Carolyn Arcure, Ravi Ravindra, Albert LaChance, Joanna Macy, Frank Cousens, Sam Keen, Patricia Mische, Dick Simpson, William MacNeill, Arthur Fabel, Peter Berg, and faculty of the Institute in Culture and Creation Spirituality, for a particularly instructive session, with Matthew Fox, Marlene DeNardo, Jim Conlon, Dody Donnelley, Neil Douglas-Klotz, Victor Lewis, Adrianna Diaz, Jeremy Taylor, Betty McAfee, Jose Hobday, Paula Koepke, Bob Frager, Elan Shapiro, Marilyn Goddard, Shanja Kirstann, Betsy Rose, Robert Rice, Kamae Miller, and Barry Gill.

One of the most rewarding moments in the whole ten-year process was the center's first symposium, where cosmologically oriented thinkers gathered together to reflect upon the strong and weak points of the narrative in our manuscript. It was exciting and deeply moving to see the story come alive in a gathering of people with such radically different perspectives. Throughout the symposium I had the sense that participating in a community's search for a common wisdom is one of the central and satisfying activities humans engage in.

We remember our symposium participants Mary Evelyn Tucker, John Grim, David Griffin, Charlene Spretnak, Bruce Bochte, Michael Zimmerman, Matthew Fox, Marnie Muller, David Peat, Duane Elgin, Ralph Metzner, John Broomfield, Mary Leahy, Bill Free, Billy Holliday, Joe Meeker, Jean Lanier, Arne Naess, Ty Cashman, Ann Jacqua, Sidney Lanier, Betty Roszak, Bill Keepin, JoAnn McAllister, Eleanor Anderson, Rex Weyler, Stuart Brown, Jurgen Kremer, Sandra Lewis, Lee Henderson, Ted Roszak, Sheila Gibson, George Sessions. Our gratitude to the organizer of the event, Sue Espinosa, to Tom Espinosa, and to our hosts at the Santa Sabina Center, Susannah Malarkey and Harriet Hope.

As we came closer to the finished manuscript, the need for competent assistance ballooned. Eileen Doyle, Karen Jones, and Anita Fasnaught did extensive preliminary work in typing the manuscript. Visual researcher Lynne dal Poggetto, assisted by Bruce Bochte and Sharon Kehoe, procured the photographs for the manuscript; while Linda Bochte transformed audiotapes and layer after layer of edited manuscripts into coherent typeface. Ned Leavitt dealt with all contractual negotiations and brought us to a happy and congenial home with Tom Grady, Robin Seaman, Kevin Bentley, and their associates at Harper San Francisco.

ACKNOWLEDGMENTS

A special thanks to Lavinia Currier, who enabled Thomas and me to spend some time overlooking the Pacific Ocean to reflect upon the manuscript as a whole, especially in light of the discussions at our symposium; and to Bruce Bochte, who, by transcribing the many large and small group discussions at the symposium, provided us with a permanent record with which to consult; and to Denise Swimme, whose unwobbling encouragement and wise counsel provided an enabling psychic energy throughout the ten years of research and writing.

Brian Swimme

The quotation from Black Elk on page 265 is taken from the book by John G. Neihardt, *Black Elk Speaks* (New York: Pocket Books, 1959).

Index